Yolanda Zappaterra

A ROTOVISION BOOK
Published and distributed by RotoVision SA
Rue du Bugnon 7
CH-1299 Crans-Près-Céligny
Switzerland

RotoVision SA, Sales & Production Office
Sheridan House, 112–116A Western Road
Hove, East Sussex BN3 1DD, UK

Tel: +44 (0) 1273 72 72 68
Fax: +44 (0) 1273 72 72 69
E-mail: mail@rotovision.com

Distributed to the trade in the United States by
Watson-Guptill Publications
1515 Broadway
New York, NY 10036

1 3 5 7 9 10 8 6 4 2

ISBN 2-88046-330-0

Book design by The Design Revolution

Production and separations in Singapore by ProVision Pte. Ltd.

Tel: +65 334 7720 Fax: +65 334 7721

CONTENTS

When I started researching this book, searching for the best in contemporary electronic illustration — looking in magazines, books, on the Internet, at advertising material and merchandising, at animation, publicity, and in comics to name a few — it seemed an easy task. Having decided that anything produced on a computer or utilising a computer fell into the remit of 'digital illustration', it quickly became apparent that while there is much work that fulfils those criteria, little of it seems to make use of the remarkable opportunities the computer affords the creative mind.

The best of this work takes illustration beyond its previous boundaries and affords the 'illustrator' much greater control and space for experimentation, while also allowing them to use different elements to arrive at something much broader than illustration but still meeting its essential meaning: pictorial symbols which act as emotional expression, warning, greeting — all forms of notification in one way or another.

The very way in which the word 'illustrate' is used — to illustrate a point — encompasses what's at the very heart of communication, the need to get a message across to people who may have different languages, customs, social mores and understanding. Good illustration can transcend all these things in a way written communication could never possibly do.

For a long time the work of commercial illustrators has been viewed as peripheral — something to break the monotony of words filling the page. While cartoons and their ability to communicate satire, irony, humour and dissent have always been rightly appreciated and valued, illustration has had to fight to be seen as more than decoration.

While many illustrators are still regarded as mere cyphers for the art editor, designer and producer, trusted to do little more than interpret a specific brief, some have gone much further than that. With the aid of the computer these creatives are pushing them-selves and their work into more coherent, completed projects, because the computer and its range of artistic software not only allows them to experiment with practical things like technique, texture and colour in ways which, pre-computer, would have cost them a fortune in time and materials, but more importantly it's allowing them to have a much greater part in the finished project or even take control of it.

The best of them — and I believe 16 of them are gathered here — have integrated existing skills and learnt new ones, often through experimentation, to arrive at something which they all insist has to be led by creativity, not by the machinery. So they mix traditional photography or acrylic painting with software such as Photoshop and Illustrator, they layer and juxtapose much as they would have pre-computer, and at the heart is always the creative process. The computer can enable them to realise ideas that may have been impossible previously, but it is simply a sophisticated tool.

Armed with these new skills and tools for greater creative experimentation, they're pushing the boundaries of illustration to a point where the term 'illustrator' becomes meaningless. As much as designers they are becoming 'visual communicators', adept at working in different contexts and media, able to realise the brief from the client in much broader terms than was previously possible or, from the point of view of many art editors, even desirable. Just as DTP has enabled anyone to potentially become a publisher, so the various layout, manipulation and art programs and plug-ins now available to anyone of a creative bent have given visual communicators access to all the tools they need to realise a creative brief in its entirety.

Some of the creatives in this book work in interactive media and have enthusiastically embraced the technical and creative opportunities offered by the web and CD-ROMs. Others work on film, video and print, creating advertising, publicity, books, kitchen-ware, packaging, models, and logos. Whatever they apply their work to, what they all share is an astonishing ability to contextualise that work, to concisely apply the thought process appropriate to the medium and project in hand. It's this contextualisation and creative adaptability which this book focuses on, rather than competence with a computer, allowing us to discover why, as much as how, a creative arrived at a particular visual solution.

Berkshire-born **DAVE McKEAN** has been working for over a decade as an illustrator. He is probably best known for his comic books: the Batman novel *Arkham Asylum* (still the single most successful graphic novel ever published); the *Sandman* series covers; *Black Orchid*; his collaborations with writer Neil Gaiman on *Violent Cases*, *Signal to Noise* and *Mr Punch*; and his own book as writer and artist *Cages*. He has also produced more than 90 CD covers for artists which include Tori Amos and Michael Nyman, and his talents extend professionally into the fields of photography, painting, sculpture, music, multi-media CD-ROMs (for the Residents and the Rolling Stones) and most recently, film and video. His clients include Sony, Kodak and DC Comics and he is a regular contributor to the *New Yorker* magazine.

McKean's work is as eclectic stylistically as it is physically. His images are assemblages of photographs, drawings, paintings and sculptures made and constructed specifically for the final piece, a process which reflects his influences; among them artists Joseph Cornell, Kurt Schwitters and photographer Joel-Peter Witkin. McKean's early collages were achieved with photographs, spray-mount and a scalpel: 'Initially they were just photographic things I could do with double exposures and objects laid on glass. I originally bought a computer because I needed access to the translucency you get with Photoshop layers,' he says.

'I LOVE COMPUTERS FOR COMPOSITING AND FOR CONTROL BUT NOT FOR IMITATING NATURAL MEDIA, SO I PHOTOGRAPH AND SCAN IN REAL OBJECTS.'

VERTIGO GALLERY

Vertigo Gallery incorporates photography, painting and sculpture. The image was composited together in Photoshop. 'When the elements were sharpened the edges and little pieces of paint were picked out and highlighted but some areas stayed slightly soft, and I like the mix.'

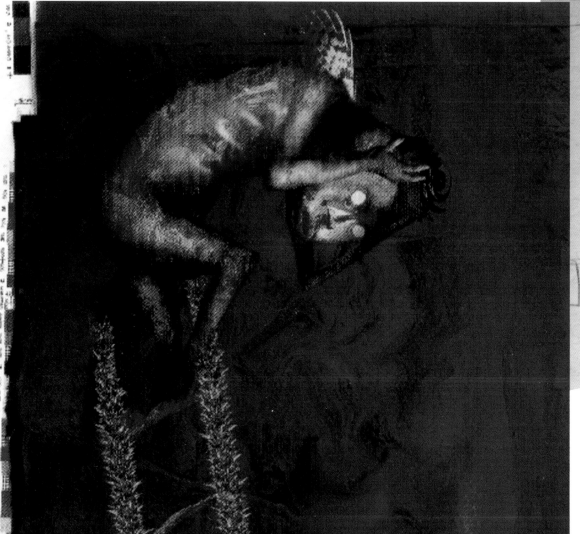

SKETCHES

'I work out the ideas for the images in notebooks, doing very "doodley" sketches. Once I've a sketch that is right I try and go with that. The first little thing that works often holds the basic design and composition, and that's where the strength of the piece lives.'

For eight years McKean has designed and created covers for Neil Gaiman's hugely popular comic series *Sandman*. *Sandman* is an extraordinary tale and is a million miles away from the usual fare of flying muscles in tights: 'the series is literary-based, the characters are devices to deal with myth, dream and established story tropes.

The main character is really just an idea given a body to inhabit, he is an archetype, as are his family, so it's an ideas book. That's why the covers ended up being symbolic rather than a literal interpretation based on a character or panel from inside,' says McKean.

Dust Covers is a compendium of all 75 covers for the *Sandman*, printed in full colour and running to 208 pages, all designed and laid out by McKean. Also included are 20 collected book covers, several special edition covers, a variety of supporting and personal images, with biographical information and annotations throughout by Gaiman and McKean. Created especially for the book is the very last story of the *Sandman*, written by Gaiman and drawn by McKean.

(the project)
THE PROJECT ⟩

SCANS ⟨ 'The rose branches were taken from the garden and photographed. The little sculpture is made out of clay and photographed but the only bit that survived in the final image is the rib cage. The rest is a painting and little bits of abstract texture.' 6 x 8 prints and the A4 artwork were all scanned in as large as possible on a desktop scanner, sharpened then reduced to the current size.

(development) DEVELOPMENT

'I've been working for DC so long, they basically trust me and just leave me alone to get on with it. Almost all my work follows the same path. First I start drawing, I work out the ideas for the images in notebooks, doing very "doodley" sketches. After that I make the media or paint the objects I need to build the image. Most of the materials are either shot or made by myself and then photographed, but very occasionally I use an old family album which is great for odd people in the background or a façade of a building.'

'Everything is scanned into the computer on either a desktop or drum scanner. Then I just begin compositing, which is usually a linear task but everything changes as it all comes together. I try and keep the basic thrust of the sketch. I allow myself any amount of play with an image but if the work starts to go off the rails then I'll rework from previously saved versions, but sometimes it means going back to the original sketch and remaking elements.'

'I have no rules, but I don't like relying on a computer — you can easily end up with this obviously slick and plastic-looking digital image with no dirt or real textures and no surprises. If you want to accomplish a task then computers are great, but you can't hold a conversation with them, they're all too clean and logical.'

PHOTOGRAPHY{ Elements in the final image include photographs of books, a friend of McKean's sitting at a table, and a blue curtain, all shown here. Moroccan carpet, Thai silk and a piece of sheet metal produced by a friend of McKean's who makes iron works also formed part of the image. All of these component parts were shot on to 6 x 8 transparencies then drum scanned.

FIRST VERSION{ This initial version was created for advance solicitation of the book to distributors. McKean then decided to change the background for the actual book cover.

REFERENCE

'The *Dust Cover* cover and back image was based on the Arcimboldo painting "Librarian". The series is literary-based and the *Sandman*'s library of unwritten books features throughout the story, so it seemed an appropriate reference.'

'THE MORE YOU WORK INTO A PAINTING WITH PAINT, COLLAGE, SOIL OR VARNISH, OFTEN THE MORE BEAUTIFUL IT BECOMES WITH LAYERS OF TIME. WHEREAS THE MORE YOU APPLY PHOTOSHOP FILTERS OR BRUSHES TO AN IMAGE THE MORE IT LOOKS LIKE A PIECE OF PLASTIC, IT LOSES ITS HUMANITY.'

(technique) and KIT ᵥ

'I always work on images in RGB and convert to CMYK at the end. I don't colour correct, I can recognise the luminous buzz on a colour that won't reproduce. I'm also used to colours going really dull on the screen when converted, like a burning red, but the information is still 100 per cent yellow and 100 per cent magenta and when it's printed the colour comes back. It took a few jobs to get used to the vagaries of it. The other thing is getting used to resolution, seeing an image on screen as either close-up and sharp, or complete but fuzzy.'

PHOTOGRAPHY

Elements of the panels photographed by McKean.

'I supplied film run-out from digital files to the printers. All images were prepared as CMYK EPS files in Photoshop 3. The design was completed as readers flats in FreeHand 5. I never use Quark, I don't like it. It doesn't feel very natural to me. It seems to be a very technical program, there seems to be several layers between my intuitive feel for what I want to do and then getting there, whereas version 3 of FreeHand feels very raw and rapid. The upgrades from version 3 have ruined a very simple and usable program. When I finish the design I open it in FreeHand 5 only to check that it will print right.'

'I LOVE COMICS AND I LOVE STORYTELLING, AND I LIKE TRYING TO FIND USES FOR WHATEVER MATERIALS I AM WORKING WITH, TO TELL STORIES IN UNIQUE WAYS. THE STORY IS NEIL GAIMAN'S LAST LITTLE SOLILOQUY ABOUT WRITING *SANDMAN* AND THE POINT AT WHERE HIS CREATIONS INFILTRATE THE REAL WORLD. THE COMPUTER ALLOWS SEAMLESS MIXES OF FACT AND FICTION, THE KEY THEME IN THE STORY.'

THE LAST SANDMAN

'The story is told in simple two-panel pages so they relate to the illustrations in the rest of the book, having something of the feeling of an illustrated children's book. The images are composites of drawings, type, abstract painted backgrounds, paper collage, and photographs.'

'YOU CAN OFTEN SEE WHERE A LOT OF DESIGNERS HAVE JUST SETTLED FOR AN IMAGE THAT LOOKS GREAT ON A MONITOR BUT LOOKS MUDDY AND OUT OF FOCUS WHEN IT'S PRINTED UP.'

HARDWARE

> Mac Quadra 950 2.4GB HD/84MB RAM
> PowerMac 9500/350 3GBHD/130MB RAM
> 20" and 21" screens
> Mirror scanner, Saphir scanner

> A3 and A4 Wacom tablets
> 88MB Syquest
> Plasmon and Yamaha CD writers
> LaserWriter Pro B/W printer for proof reading

SOFTWARE

> Freehand 3 and 5
> Photoshop 3
> Painter (used once, never again)

DAVE McKEAN
FAX: +44 (0)1797 270 030

CLIENTS INCLUDE: DC Comics; Virgin Records;
Saatchi & Saatchi US; SONY; *New Yorker*. [SEE PAGE 145]

Clau

In the four years since **Claudia Newell** took up professional illustration, the Portland-born artist has worked for a wide variety of clients, ranging from esoteric titles such as *Guitar Player* and *Urban Outfitters' Slant* to more traditional journals such as *The Village Voice* and the *Los Angeles Times* magazine. She's also worked for TV companies Nickelodeon and Bravo, and record company Hemiola Records.

Newell's mix of diagrammatic style with bold colours is inspired by a variety of disciplines and techniques: 'Asian Food packaging! Comforting Sanrio-type simple children's characters – I've always been given a lot of dry "cyber" subjects, so I try to make them friendly. I also love any kind of instructional diagram, whose function generates its form. 1950s cartoon characters. 1950s and 1960s paperback book covers, where psychological difficulties and philosophical systems are represented by geometric shapes,' she enthuses.

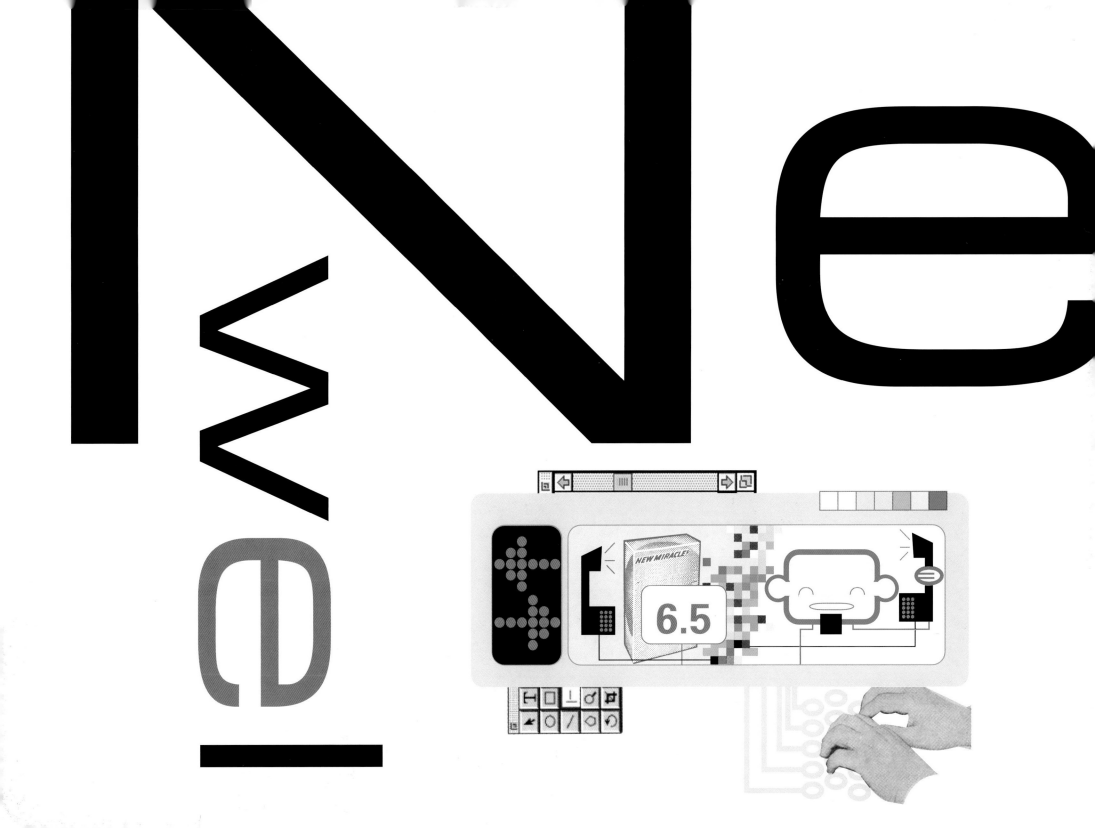

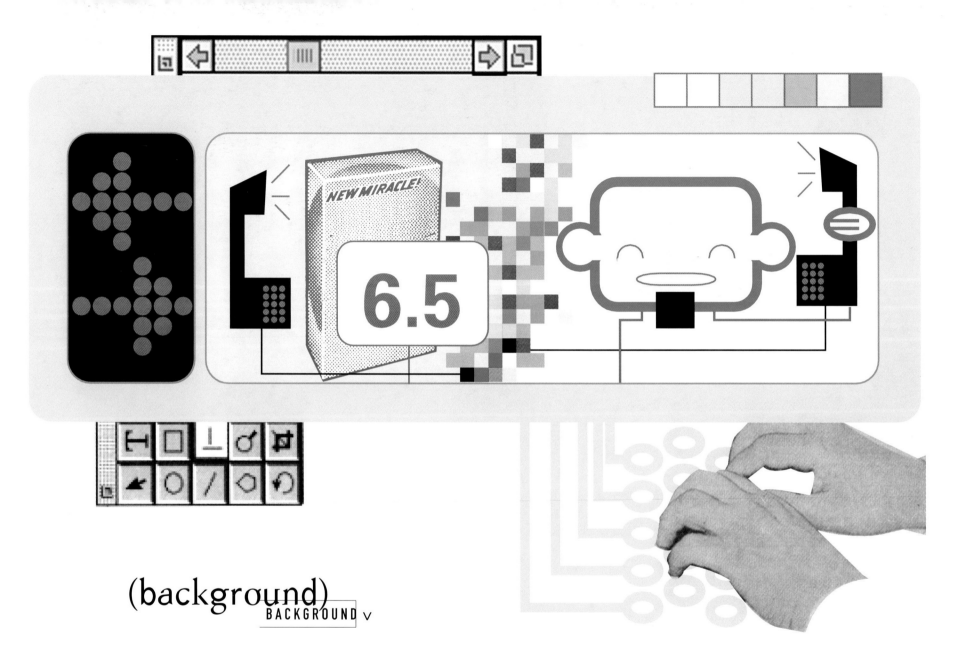

NEW MIRACLE!

6.5

(background)
BACKGROUND ∨

When software company Adobe launched its DTP package PageMaker 6.5, *Adobe Magazine*, the company's journal for registered users of its products, ran an article about the package's scripting features. These cover fairly abstract notions which are best described as 'conversing with it via scripting to accomplish desired goals', says Newell.

'The people at Adobe had no particular idea of what they wanted, except for the number of illustrations – three – and specific points they were commenting on. So as this is pretty dry subject matter, with no tangible elements, I decided to personify it through these little characters and try to make it come alive, get to the core of the ideas that were being expressed and exaggerate them,' explains Newell of her commission.

MAIN IMAGE
The typing hands, scanned in from a 1930s business school instruction manual, were vignetted by tracing the outline in Illustrator and masked with that shape.

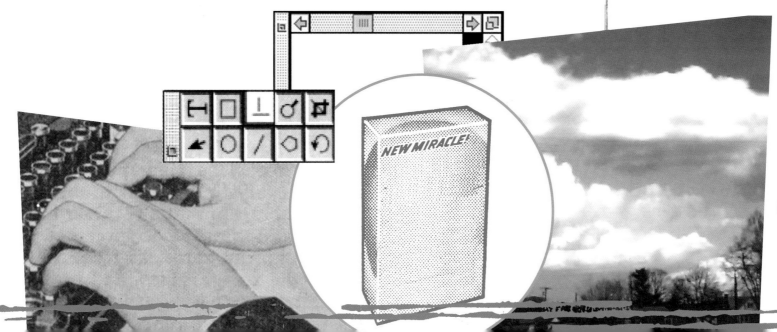

(the project)

THE PROJECT ᵥ

As the article's author Olav Kvern had mentioned that PageMaker 6.5's scripting ability could be compared to a conversation, Newell decided to take this idea as the basis of the three illustrations for the piece, one main one and two smaller ones. These three aren't so much in chronological order as illustrating different aspects of the article, so Newell had to tie them together but give each its own storyline.

The main illustration (facing page) uses the conversation theme most strongly, while the second one relates to the author's discussion of 'conditions' in scripting and the final one illustrates the idea of scripting as delivering commands to the program.

TECHNIQUE

Some of the appropriated images were brought in via scanner: the box is from an old letterpress/ metal-cut printed supermarket ad. This was scanned in at a high resolution to preserve the dots, which made an interesting contrast to the flatter shapes in the drawing. In some cases, Newell masked it inside a drawing of its own outline to make it stand out against a contrasting background. She also changed its colour from black to the dark blue/green using Photoshop.

COMBINATIONS

Much of Newell's work is created from combinations of very simple shapes, as these screenshots, preview and keyline views show. 'The pixel-y squares were nabbed from some other experimental drawing I did where I used a Mosaic-type plug-in in Illustrator on a random placed image to generate a bunch of squares,' explains Newell.

'I usually start by reading the thing and considering anything which may have been said by the art director. I then make a bunch of tiny pen sketches just for myself to get out what I'm thinking. I tend to have too many ideas at first and need to prune them back into a clearer statement. Then I start drawing in Illustrator, with simple shapes,' explains Newell. 'I usually have some sort of vague mental inventory of appropriated images I've collected which may resonate with a given subject matter/assignment as well,' she adds.

'PM 6.5 is a pretty abstract product, it's not a blender or a coat or whatever, so I decided to use a box illustration lifted from an ageing newspaper to guest star as and personify PageMaker. Since all this scripting takes place by the user typing in commands/scripting language, I had the typing hands and abstracted keyboard with the above conversation placed within the yellow, vaguely monitor-like shape: accessing this cyber communications world via our typing interface. Lastly I took some screen shots of the PageMaker tool box and window edge, because they look good and are part of the environment of our story,' says Newell.

DEVELOPMENT
(development)

The second illustration deals with an idea the author discussed about 'conditions' in the scripting. 'I believe he wrote "If it's raining, then bring an umbrella," so a huge thanks to him for giving me this easy way to express the concept!' says Newell. So here the umbrella personifies PageMaker, shielding the user. 'I put in a GATF registration-type symbol in as the obscured sun behind the clouds for fun, and after all, all this scripting is going to affect what one is preparing to print.'

The final illustration (overleaf) had to express how scripting is basically telling PM to carry out tasks. 'For some reason this brought to mind the idea of a surgery with the Doctor saying "scalpel" and Nurse responding "scalpel".' It's a straightforward idea represented by documents from the documents palette as the patient, with the box (PageMaker) and the doctor (scripting) combining forces to work on it.

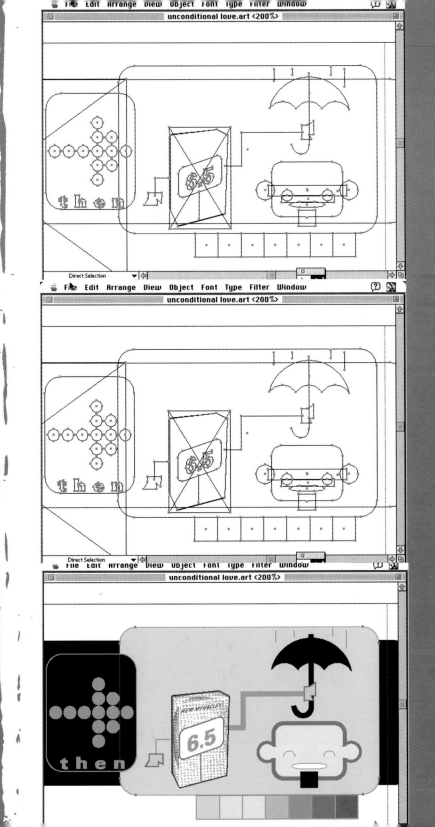

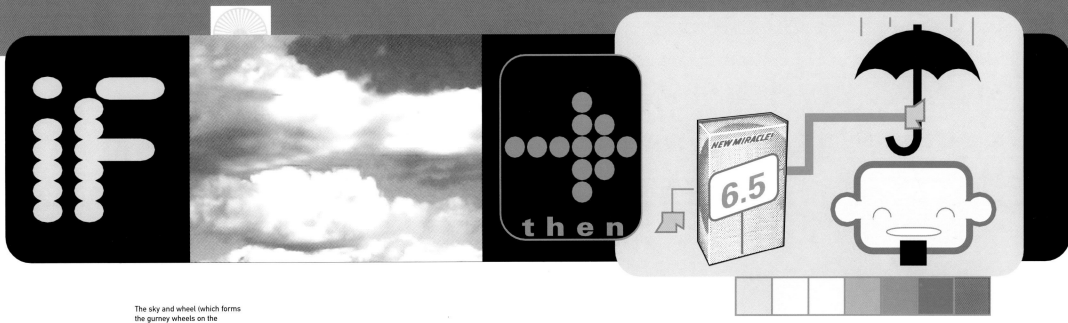

PHOTOGRAPHY

The sky and wheel (which forms the gurney wheels on the illustration overleaf) were shot using an Apple QuickTake 150 digital camera, and loaded on to the computer using the control panel software. There they were converted to black and white, and converted to the correct file format (EPS) to be placed in the illustrator drawing.

'SOMETIMES I WISH I COULD JUST TAKE A PEN AND CHANGE SOMETHING — DRAWING IS NOT VERY NATURAL IN THE VECTOR PROGRAMS. IN THE AMPLIFIED ARENA OF THE COMPUTER SCREEN IT'S EASY TO FUSS WITH ALMOST IRRELEVENT DETAILS ENDLESSLY — IT'S STILL VERY IMPORTANT TO DO A RELAXED SPONTANEOUS SKETCH FOR ONESELF WITH A PEN, AND PRINT YOUR WORK OUT TO ASSESS THE OVERALL PICTURE.'

'ILLUSTRATORS WHO HAVE A COMPUTER CAN BE PRESSURED TO CAVE IN TO THE CYBER-FAD OF THE WEEK, MARBLED THREE-DIMENSIONAL MODELLING AND SO ON. ALL THIS ABILITY TO DO THINGS HAS TO BE TEMPERED WITH SOME SORT OF EDUCATION IN GRAPHIC DESIGN, GIVEN THE ONSLAUGHT OF HORRIFIC TYPOGRAPHY AND PROOFREADING ERRORS CONFRONTING US EVERYWHERE.'

(technique) and KIT>

The three pieces were all done in Illustrator and Photoshop, and quite old versions as the project was completed in autumn 1996. 'I wasn't doing anything too demanding or processor-intensive, luckily', says Newell. 'I generally just make everything by building with the simple geometric shapes available in the tool palette. For things that are a more free-form shape, like masking around the hands or box, I just draw around it with the pen tool,' she adds.

The various scanned elements, consisting of photos taken by Newell — such as the wheels on the patient's gurney and the storm clouds — along with images such as the box of washing powder and the hands at the typewriter, were combined with Newell's own Illustrator-created diagrams and screenshots to arrive at the finished pieces.

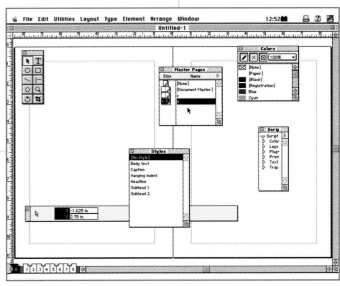

HARDWARE

> Macintosh Centris 660AV, 16Mb RAM

> 13" colour monitor

> Umax Vista Scanner

> Apple QuickTake 150 digital camera

SOFTWARE

> Adobe Illustrator 5.5

> Adobe Photoshop 3.0

CLAUDIA NEWELL
484 CAPISIC STREET, PORTLAND, MAINE 04102, USA
TEL: +1 212 969 0795
E-MAIL: CNewell937@aol.com
WEBSITE: http://www.mindspring.com/~cnewell

CLIENTS INCLUDE: Nickelodeon; *The Houston Times*; *Los Angeles Times* magazine; Hemiola Records; *The Village Voice*. [SEE PAGE 144]

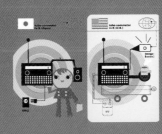

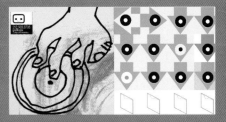

While the world wide web is
awash with illustrators' portfolios
and homepages, it's fair to say
that they've failed to capitalise on
the web's opportunities for
creative expression in the way
graphic designers have, with
a few notable exceptions.
37-year-old West Coast illustrator
J. OTTO SEIBOLD is one such
exception, with web sites
promoting not himself, but two
wonderful characters called
Mr Lunch and Space Monkey,
who also feature in books and on
T-shirts illustrated by Seibold
and written by his wife and
co-writer Vivian Walsh.

'I NEVER WENT TO **ART SCHOOL** AND CONSEQUENTLY I NEVER STOP PURSUING THE THINGS I LIKE. THE ONLY DISCIPLINE I NEEDED WAS THE ONE THAT LET ME CONTINUE TO **LEARN**. I WANT TO FEED MY HEAD UNTIL I AM DEAD.'

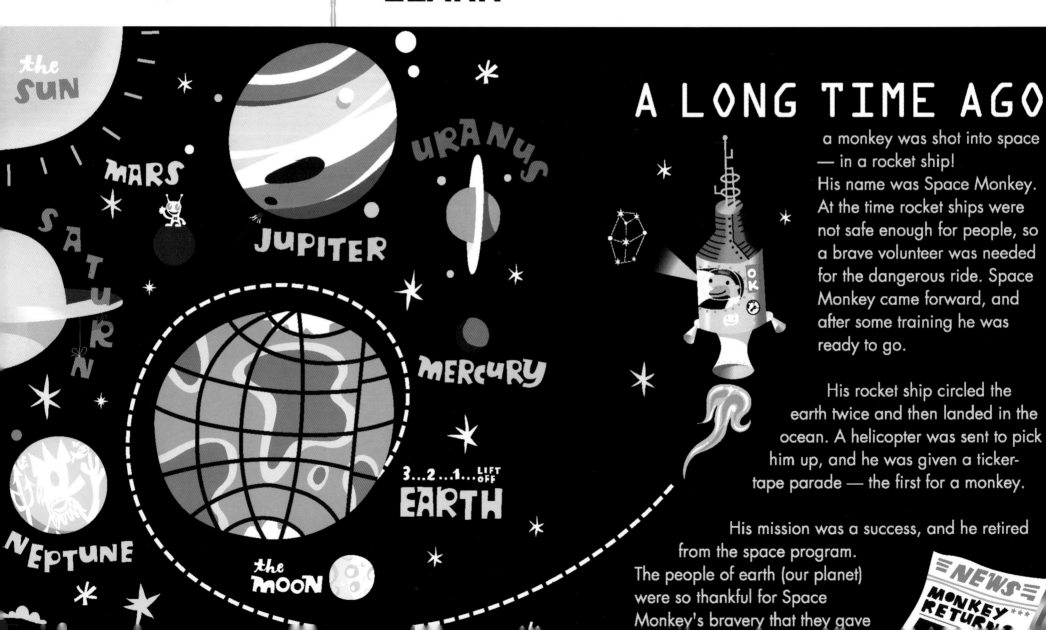

A LONG TIME AGO

a monkey was shot into space — in a rocket ship! His name was Space Monkey. At the time rocket ships were not safe enough for people, so a brave volunteer was needed for the dangerous ride. Space Monkey came forward, and after some training he was ready to go.

His rocket ship circled the earth twice and then landed in the ocean. A helicopter was sent to pick him up, and he was given a ticker-tape parade — the first for a monkey.

His mission was a success, and he retired from the space program. The people of earth (our planet) were so thankful for Space Monkey's bravery that they gave

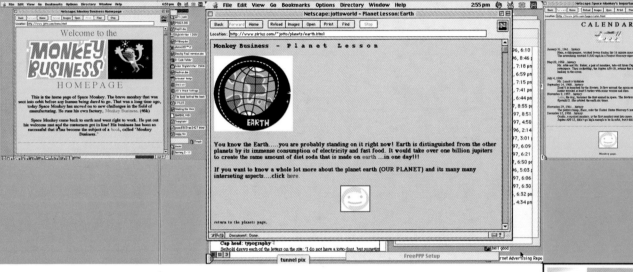

INFORMATION While Space Monkey is obviously fictitious, his site offers a great deal of information and links to educational sites in a witty and accessible way.

spacemonkey, CEO
monkey business

THE WEB AS PROMOTION

Space Monkey is a successful ex-astronaut who is now a successful businessman, and as such obviously early on saw the potential and marketing importance of a website: 'He's a desperate self-promoter, and the promotion is the content in a way,' says Seibold.

With no formal training in art but having done drafting classes at high school, Seibold entered the workplace as a draftsman, drifting from corporation to corporation before arriving at the design and research unit of the Clorox Corporation. Here he had not only access to the most powerful computers of the time, but was also encouraged to explore creative outlets. Having become interested in 'drawing for a living', he started by designing and illustrating the company's calendar, which led to the creation of a portfolio. 'I began by cutting shapes from coloured paper and gluing them together, but was then invited to spend some time with Adobe learning how to use Illustrator,' he recalls. 'While initially the process was the same, plotting colour shapes on screen and applying them next to or on top of each other, it did get rid of handicaps like having to mix paints,' he adds.

Seibold began using a computer in 1987, having previously bought a Macintosh 128, which he describes as 'etch-a-sketchy, not something to make pictures on', when it came out in 1984. Work as a commercial artist quickly took off: 'I have worked with many corporate giants and many established periodicals... their names blur together'. His reticence to name names is, I suspect, something to do with the fact that with the help of the world wide web, he and Vivian are now close to having a cottage industry which is in the enviable and rare position of not needing conventional clients. Instead, they generate books, T-shirts and sites which satisfy their own creative needs without having to kowtow to the needs of clients. 'I still do commercial art, if the job is interesting, it's a good way to keep involved with more engaging projects,' says Seibold.

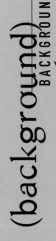

'J. OTTO SEIBOLD DRAWS ALL OF THE TIME.
IT'S HIS JOB.
HE'S A PROFESSIONAL.
HE DRAWS ON A COMPUTER WHICH MAKES
HIM SORT OF LIKE A SCIENTIST.
BUT HE DOESN'T WAKE UP UNTIL NOON,
WHICH MAKES HIM LIKE AN ARTIST.'

(the project) THE PROJECT

The decision to give Space Monkey a website in 1995 seemed a natural one: 'We saw the web page initially as an extension of the premise we had in the book... the monkey had a factory and so naturally the monkey had to have a website... I think our first attempt at making the proper html code was copying the Mobil oil homepage and substituting our monkey CEO picture for theirs. It evolved from there,' says Seibold.

SKETCHES
J. Otto still does pencil sketches of characters and scenes before creating them fully in Illustrator, and he also draws 'little octopus drawings to show how pages might link together'.

(background) BACKGROUND

The 20th anniversary of the first moon walk prompted Seibold and Walsh to look back over those years and think 'where next?'. So they created the character Space Monkey, the first monkey in space and a successful astronaut who settles down to a life of entrepreneurial bliss, with his own factory and a close friend, Penelope the Bug. Set in the not-too-distant-future, *Monkey Business*, a 32-page children's book was born, and as with the previous *Mr Lunch* books, drew comparisons with the work of Richard Scarry, who Seibold and Walsh make knowing references to in those first books.

CHARACTER DEVELOPMENT

Penelope the bug is firstly an enemy then a friend of Space Monkey's: 'The personality of the characters comes across really well on the website,' enthuses Vivian Walsh, co-writer and coder of the site.

REALISM

While the Space Monkey is fictitious, the site to all intents and purposes acts as a real site, perfectly aping the millions of promotional sites on the web, complete with advertising banners.

www.
jotto
.com

Monk E.business

DEVELOPMENT
(development)

'In the early days of the website, we worked with several people, but now Vivian does the html coding and the site is updated frequently,' says Seibold. The site acts more than anything as a promotional area for the books and T-shirts, but does so in a subtle and witty way: 'We wanted the promotional aspects to be far in the background of our all-new content... but the fact that the fictional monkey was a desperate self-promoter helped blend these delicate concerns... the promotion is the content in a way,' explains Seibold.

A popular area of Space Monkey and the Mr Lunch sites is the contact and feedback one, which receives hundreds of messages from both adults and children who give ideas, send in drawings of the characters, ask questions and offer praise and encouragement on all areas of the Seibold/Walsh output. All the mail is put up on the site and is responded to by Vivian — obviously in the guise of Space Monkey.

'What I particularly like about the Internet is the fact that it gives us more room to play with our characters than the confines of a book could. We can put narrative into it, flesh out the characters so their personality comes across better and include the written and drawn feedback, which is really popular and simply couldn't be done in the books,' enthuses Vivian.

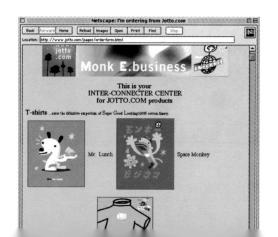

FEEDBACK { The feedback part of the site is extremely popular, allowing fans to post drawings and comments on to the site. All receive a response from the monkey (a.k.a Vivian). Visitors can also purchase books and T-shirts at the site.

the site: 'I do not have a jotto-font, but sometimes I cheat and copy some letters from an old work… but not that much,' he admits.

(technique) and KIT ᵥ

'The website would be a perfect job for us, if only you could make money from it. But as we use it for experiments, feedback, contacts, promotion and marketing we keep it going,' says Seibold. 'But as we can't count on it as a source of income, we make most of our decisions while doing something else… like at meals… we talk with food in our mouths. Vivian does all the html coding and most of the words, and answers most of the letters, while I make all the images and come up with some of the concepts. I draw little octopus diagrams to show how pages might link together and we sit in front of the computer together now and then. That is when most good stuff gets done,' says Seibold, who creates all the imagery using Adobe Illustrator.

The Space Monkey and Mr Lunch sites are fairly straightforward, allowing anyone who visits the sites the same experience regardless of their browser and plug-ins. Both sites use small amounts of animation. Space Monkey has the garden scene shown here while Mr Lunch, the professional bird chaser, uses a loop: 'Mr Lunch running in place is an animation loop that seems just right for a web page… I often get the feeling when using the web that I am running in place too,' says Seibold.

ANIMATION

HARDWARE

> POWERPC 8500
> EPSON COLOUR PRINTER
> 'AN OLD UMAX SCANNER THAT LOOKS LIKE IT CAME OUT OF THE CENTRE CONSOLE OF A MINI-VAN.'

SOFTWARE

> ADOBE ILLUSTRATOR 6.0
> WACOM TABLET 'WITH A PIECE OF BOND PAPER TAPED TO IT. WITHOUT THE PAPER TENSION IT FELT LIKE DRAWING ON AN ICE CUBE.'

J. OTTO SEIBOLD
1261 HOWARD STREET, #3, SAN FRANCISCO
CA 94103, USA
TEL: +1 415 558 9115 FAX: +1 415 558 9131
E-MAIL: jotto@pop.sirius.com (j.otto seibold)

CLIENTS INCLUDE: Jean Paul Getty Centre; Walker Art Centre. [SEE PAGE 150]

Steven
R.

'Everything I have done for Holy Body Tattoo revolves around this simple sacred heart motif done to promote a performance entitled "Poetry and Apocalypse", and although it can change on any given project it is probably the most recognisable symbol relating to the company.' says Gilmore.

m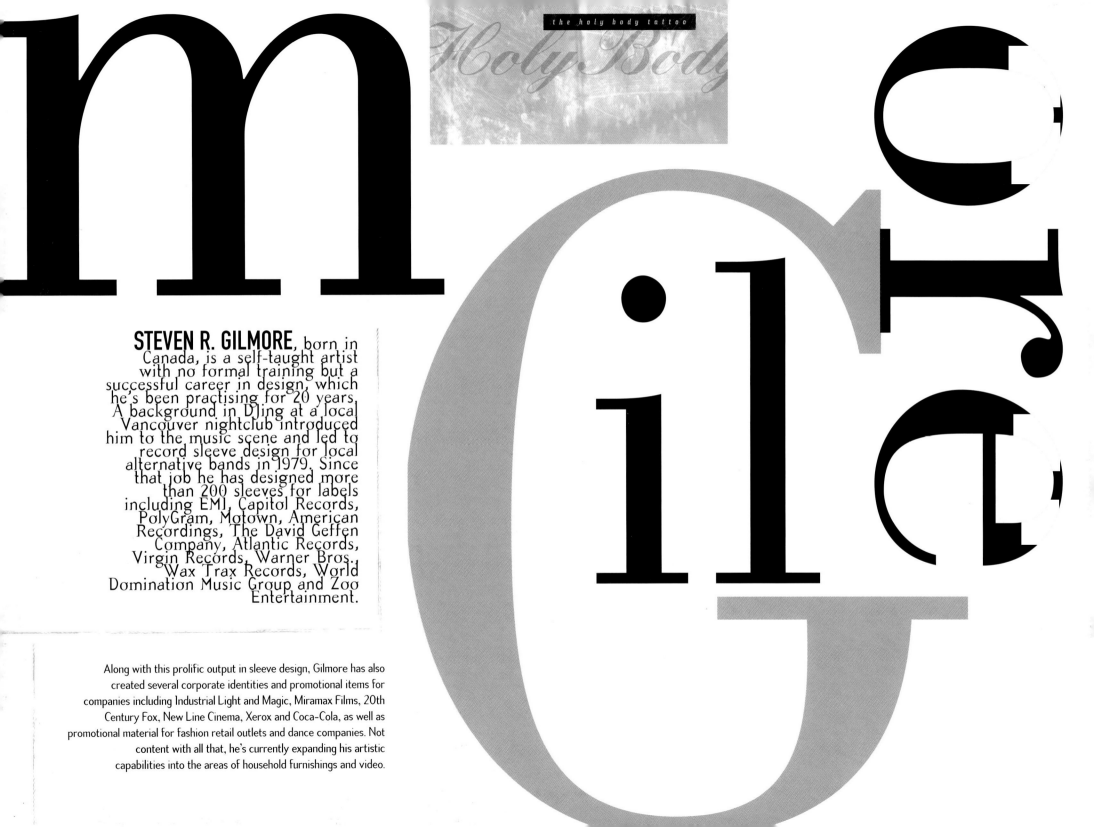

STEVEN R. GILMORE, born in Canada, is a self-taught artist with no formal training but a successful career in design, which he's been practising for 20 years. A background in DJing at a local Vancouver nightclub introduced him to the music scene and led to record sleeve design for local alternative bands in 1979. Since that job he has designed more than 200 sleeves for labels including EMI, Capitol Records, PolyGram, Motown, American Recordings, The David Geffen Company, Atlantic Records, Virgin Records, Warner Bros. Wax Trax Records, World Domination Music Group and Zoo Entertainment.

Along with this prolific output in sleeve design, Gilmore has also created several corporate identities and promotional items for companies including Industrial Light and Magic, Miramax Films, 20th Century Fox, New Line Cinema, Xerox and Coca-Cola, as well as promotional material for fashion retail outlets and dance companies. Not content with all that, he's currently expanding his artistic capabilities into the areas of household furnishings and video.

(background)
BACKGROUND
ᐯ

Canadian contemporary dance company The Holy Body Tattoo saw examples of Gilmore's work in 1994 and felt they shared a similar aesthetic sensibility with him. Keen to avoid being represented in the same way as the more traditional companies, one radical departure is the absence of any company photographs on their promotional posters, or even the use of the word 'dance'. 'The aim of their posters is to set a tone of intrigue; unless you are familiar with the company you wouldn't know whether it is a dance, theatrical or musical performance,' says Gilmore.

'When I first started working with the company I didn't have a clue about the dance community and how to approach it. I hadn't seen any of their performances so the only thing I had to work from was my interpretation of their name and the title of the piece they wanted me to work on, "Poetry And Apocalypse". The one solid direction I had was that they felt the "Sacred Heart" captured the essence of the company; a notion that all experiences in life leave a mark on both the body and the spirit, which are sacred,' explains Gilmore.

Gilmore's initial work on the publicity material for 'Poetry and Apocalypse' led to a long-term relationship with the company which has now seen him create not only posters and related materials for individual shows, but also all the company's business collateral as well. When asked to participate in this book, rather than look at one piece of work, Gilmore opted to include the whole canon of work for the company, seeing it very much as one body of work which is constantly evolving and extending.

Unusually, he works very closely with the company in order to represent them in a way they can both feel comfortable with: 'When we first started working together there wasn't a lot of interaction between The Holy Body Tattoo and myself with regard to the creative process of my interpretations of their ideals. But over the years our collaboration has deepened and become more specific through our working together,' says Gilmore.

(the project)
THE PROJECT

'ALMOST ALL OF THE WORK IS PURELY INSTINCTUAL; ALTHOUGH IT IS NOT A CASE OF "THROWING ANYTHING AGAINST THE WALL TO SEE WHAT STICKS", IT IS A PROCESS OF TRIAL AND ERROR.'

the **holy body tattoo**

FOLDER AND CONTENTS DESIGNED BY STEVEN R GILMORE

PRESS RELEASE **FOLDER**

'I may go through several ideas before I finally decide to send a client something to look at, but in the case of this press release folder, almost every idea I had beforehand seemed to work "on paper",' says Gilmore.

PRESS RELEASE **INSERTS**

Once The Holy Body Tattoo approved the basic designs, Gilmore had a large Iris print created of all the files on one sheet showing the results in colour and how the individual pieces worked together as a whole. He now sends everything over the Internet for the company's approval as full-colour JPEG files.

steven r. gilmore electronic workshop

TYPOGRAPHY

Gilmore uses a combination of hand-drawn and computer fonts. On this press release insert, the 'B' is based on a decorative letter he found in an old type specimen book. Because of its complexity it took several hours to render.

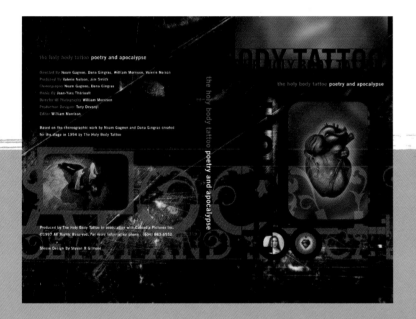

The Holy Body Tattoo

Dana Gingras, Noam Gagnon and Jean-Yves Thériault believe powerful experiences leave marks on the soul. Revisited and relived, such marks become, in effect, sacred. They are like tattoos; vivid, strange and hinting at secret stories. They cannot be stolen or erased.

"The essence (of the tattoo) evokes a poignancy unique to the mortal human condition; a true poetic creation, it's always more than meets the eye."
Modern Primitives

"The body is erotic because it is sacred"
Octavio Paz

Gilmore began by representing the 'Sacred Heart' in a non-traditional manner using acrylic on particle board. It is this illustration of the heart motif that has been featured in almost all of the company's promotional material. From there he developed the publicity material and CD booklet for 'Poetry and Apocalypse' before moving on to business cards, press packs and further performance publicity. All share the use of photography, religious imagery, found imagery and typography.

'The Holy Body Tattoo supply me with group photographs and most of the found religious images, leaving to my discretion how I would like to work with them. Since we don't use group photography on any of the major promotional pieces, they are usually manipulated to fit in with the overall feel of the specific piece I am working on.'

For the source material, Gilmore does a lot of rummaging around in thrift stores: 'A number of the images that I find inspiring or that I use directly in my work are from the books or album sleeves that I have found. I am particularly fond of old children's books plus turn-of-the-century encyclopaedias and type specimen books. I have file cabinets full of images I've cut out over the years that I can draw from,' he explains. 'The textures can be anything from a piece of distressed paper I have found on the street to a photograph that has been damaged by water. I then photograph them or directly scan them into my computer. The blurred colour fields are created basically the same way,' he adds.

 TEXTURES

Gilmore describes his work for 'Our Brief Eternity', a mixed media performance by the group, as 'a radical departure from anything I had done in the past'. The starting point was the creation of 200 slides for the performance incorporating text written for the piece by sci-fi/"cyber-punk" author William Gibson. Texture photography and design on the slides was all done by Gilmore.

DEVELOPMENT
(development)

Most of the early photography or photographic manipulation on the company's work is by photographer Anthony Artiaga. The Holy Body Tattoo had sent Gilmore a number of photographs to use on the CD sleeve for the music from 'Poetry and Apocalypse' but he felt they weren't distressed or bleak enough, so he asked Artiaga to 'work his magic on the supplied images. I was familiar with Anthony's techniques in the darkroom and I knew how capable he was of turning what seemed to be a rather ordinary photograph into something spectacular,' says Gilmore.

He uses a combination of hand-drawn type and computer fonts: 'For instance, on the first page of the press kit, the large "B" which is based on a decorative letter I had found in an old type specimen book took several hours to render, whereas most of the other type is from my library of computer fonts,' says Gilmore.

'Although I have used similar techniques and colour palettes on other projects I think, or at least I hope, that I have created a unique look for The Holy Body Tattoo,' he concludes.

'A POWERFUL COMPUTER ENABLES YOU TO BREAK FREE OF METICULOUSLY PLANNING THINGS OUT AND STILL WORRYING ABOUT BEING UNSATISFIED WITH THE FINISHED RESULTS, AND IT GIVES YOU THE FREEDOM TO QUICKLY MANIPULATE AND COMPLEXLY LAYER LARGE IMAGE FILES.'

(technique) and KIT ∨

'I use FreeHand 5.5 exclusively for rendering graphics and layouts. For example, to create multi-layered type and graphics in any given project I will first do all the layout and positioning in FreeHand 5.5 then import each layer (by exporting them as Adobe Illustrator files) into a special multi-channelled grayscale document I have created in Photoshop 2.05, these channels are then used as masks. I have kept a copy of this older version of Photoshop specifically for this purpose, you cannot type in the percentage size of the graphic you are importing with the newer versions and since a number of my graphics run off the page this feature is important to me.'

'Because the Photoshop documents can have up to 20 channels (or masks) I make sure I write down the channel number and what graphic or type it contains. I then create a new document the same pixel size as the 'mask' document, and open all the separate image (TIFF) files for a given project. After loading the selections from the 'mask' document into the new document, I then copy and paste the separate elements into these selections, after I am satisfied with the way everything is combined I save the document as one complete image (TIFF) file.'

'I would be able to save a considerable amount of time by using FreeHand's capability of pasting image (TIFF) files within a document's graphics, but this function has proven to be problematic when outputting to film so I prefer to do my final layout in Photoshop 3.5. By creating my final layout in Photoshop 3.5 I also have much tighter control over my colour palette, this way I can change the hue of a particular section directly in Photoshop instead of having to switch back and forth between two programs.'

COLOUR PALETTES ⌐

For these poster designs Gilmore drew heavily on Japanese pop culture mixed with religious and industrial imagery. 'The pastel and foreign colour palettes of Japanese packaged food were particularly inspirational. We had originally discussed using only a pastel colour palette but after I sent The Holy Body Tattoo the design they expressed a concern over it not being powerful enough for their needs, so I created a darker version as well. In the end we couldn't decide which design we thought worked the best so we used both of them.' says Gilmore.

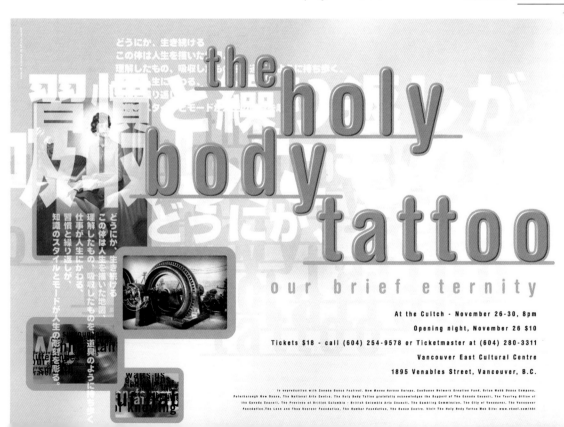

the holy body tattoo

our brief eternity

At the Cultch - November 26-30, 8pm
Opening night, November 26 $10
Tickets $18 - call (604) 254-9578 or Ticketmaster at (604) 280-3311
Vancouver East Cultural Centre
1895 Venables Street, Vancouver, B.C.

For 'Our Brief Eternity', the company wanted colour to replace the bleak monochromatic, metallic world of 'Poetry And Apocalypse'. 'That is not to say there isn't a dark edge to the performance, but the influences of that darkness are different, drawing from a number of sources; Japanese pop culture, corporate America, misinformation, miscommunication and consumerism, to name but a few. This was probably the most challenging project I have worked on to date for The Holy Body Tattoo, incorporating all these ideas into one piece of design while maintaining some kind of consistency,' says Gilmore.

JAPANESE INFLUENCES

HARDWARE

> MACINTOSH 9500/132 POWERPC (200 MEGS OF RAM)
> 17" MULTI-MEDIA APPLEVISION MONITOR
> AGFA ARCUS II SCANNER
> NIKON F-601M CAMERA AND SEVERAL DIFFERENT LENSES

SOFTWARE

> MACROMEDIA FREEHAND 5.5
> ADOBE PHOTOSHOP 2.05
> ADOBE PHOTOSHOP 3.5
> ADOBE DIMENSIONS 2.0

> ACRYLIC ON PARTICLE BOARD

STEVEN R. GILMORE
8702 CHALMERS DRIVE, LOS ANGELES, CA 90035, USA
TEL: + 1 310 659 4516 FAX: +1 310 659 4517
E-MAIL: srg@graphic.com
WEBSITE: www.webstorm.com/~srg

CLIENTS INCLUDE: EMI; The David Geffen Company; Warner Bros.; Polygram; Industrial Light and Magic; Miramax; Coca-Cola. [SEE PAGE 155]

FORK UNSPLASH

Netscape: american tourist

Back Forward Home   Reload Images Open Print Find   Stop

Location: http://www.suture.com/tourist/tour.htm

★Young Americans...   ★Yo

american
tour

click one of the sliding photos above

suture | 1 | 2 | 3 | 4 | 5 |

spoken word

a year in europe
words and images

One day I took a trip

FORK UNSTABLE MEDIA is a digital design studio housing a mixture of young American and German designers and programmers specialising in new media. The three partners – David Linderman (art/creation), Jeremy Abbett (art/creation) and Manuel Funk (marketing direction) – are backed up by four additional people supporting art, programming and contact positions: Sascha Merg ('creative' lingo programming and screen design), Christian Schaumann (project manager), Anne Eickenberg (screen design and photography) and Jan-Michael Studt (java programming).

'We have very mixed backgrounds. Jeremy is an Asian-American who spent his formative years in Texas, Manuel is from southern Germany and escaped to New York, and I left a family of religious fanatics in the heart of the American midwest,' says Linderman. The studio is based in Hamburg, Germany: 'Why Germany? Germany was a great cultural centre. Its current identity crisis and lack of direction for the next millennium is a constant source of inspiration for us. We like to play with the German identity, make fun of its mulish ways... play with its history and dig out the skeletons in the closet. We see Germany as a playground for pioneering in new media. America has an established community of pioneers, but in Germany there's an open frontier, but no homesteaders,' says Linderman.

(background)

BACKGROUND v

Projects range from web- and trash-game design, CD-ROM and diskette promotions to flyer graphics and video presentations. 'Skate culture, biking, music and clubbing help kill the pain,' says Linderman. Clients include Beiersdorf AG (makers of Nivea skin products), www.nivea.com; b&d Verlag (publisher of *Snowboarder*, *Inline Skater*, *Freedom BMX* and *Surfer* magazines in Germany), www.blondmag.com; Spar Deutschland (largest grocery chain in Germany), www.spar.de; Hamburger Sparkasse (German bank); Astra Brauerei (brewer); and Container Records, Hamburg (electronic music distributor).

(the project)

THE PROJECT v

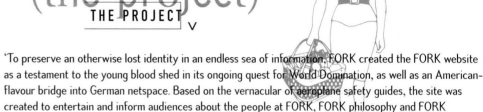

'To preserve an otherwise lost identity in an endless sea of information, FORK created the FORK website as a testament to the young blood shed in its ongoing quest for World Domination, as well as an American-flavour bridge into German netspace. Based on the vernacular of aeroplane safety guides, the site was created to entertain and inform audiences about the people at FORK, FORK philosophy and FORK projects and clients,' says Linderman.

'It is constantly updated with our latest sketches of ideas, games and graphic explorations. FORK uses safety glyphs familiar to air travel in addition to sound and games as site interface. Navigation is the experience. Sound is the ambience. Glyphs (or iconography) with new context entertain through irony. Some fun projects of late: Princess Diana Tunnel-Racer game (featuring the A-Klasse from Mercedes); Helmut Kohl as Tomagotchi-Sumo; Bad-Service sticker campaign (implementation in various Hamburg and Cologne stores, banks and Telecom phone booths).'

'WE LIKE TO PLAY WITH THE STEREOTYPE OF GERMANY. WHAT DO YOU THINK OF WHEN YOU HEAR THE WORD GERMANY? WE THINK OF A BUNCH OF DARK AND LOFTY ATTEMPTS AT WORLD DOMINATION MIXED WITH AN OVER-RIPE HISTORY AND ROMPING BIERGARTENS... (YEAH AND BAUHAUS).'

BLOND MAGAZINE 2.0

TECHNIQUE { 'We take pictures and video, sound bites and all other things that happen which strike us as being worthwhile and try to use them in our work. That way it's more personal, has more depth, and is not so generic, like... stock photography,' says Abbett.

Back | Forward | Home | Reload | Images | Open | Print | Find | Stop

cation: http://www.fork.de/home.htm

  HOME GAMES CLIENTS OXYGEN CONTACT LUFT BRÜCKE

HOME

LUFTBRÜCKE

INDEX

!!! our very new x-mas special !!!

STYLE FROM THE STATE!

QUALITY CONTROL

4RK   4RK

Fork Unstable Media
Juliusstrasse 25
22769 Hamburg
Bundesrepublik Deutschland

fon' +49 (40) 432 948 - 0
fax' +49 (40) 432 948 - 11
e.mail' info@fork.de **or** here

DANKE

Für Ihre Sicherheit
For your safety

B 737
B 737-200/300/500

Bitte Sicherheitsinstruktionen nicht von Bord nehmen
Please do not remove this card from the aircraft

Lufthansa

OXYGEN

INSPIRATION

Aeroplane safety guides were the main inspiration behind the FORK site. 'As Americans, we constantly find details in German culture to bitch about. The over-riding graphic theme of the FORK site is the crash – an answer to the most popular question (why did you move to Germany?). We wanted to describe it in the most diplomatic way possible: it was an accident. The icons on the left side of the instruction sheet are pure candy,' explains Linderman.

CREATING A GRAPHIC VOICE

'Signage and artefacts from the 1950s or 1960s are easily recognised (or reproduced) through colour, text and photographic or illustrative treatment. Borrowing this vernacular and changing the context is one way to create a graphic voice or flavour…'

...AND FLAVOUR

'…Making a tribute to the late Princess Diana with a Tunnel-Racer game is another.'

A-KLASSE

The FORK website began in the summer of 1996. Since then it has had three redesigns, taking on the form shown here in January of 1997. 'We're constantly experimenting with ideas on the site: voting for a favourite Charlie's Angel, Helmut Kohl or Sascha (our lingo king), in addition to archiving Shockwave sketches or building a new kill game on weekends or during downtime,' says Linderman: 'We try to add new games or editorial ideas about once a month — usually in the game section or special links through the front page,' he adds, continuing 'it's become pretty personal. The site helped to create a dialogue with other designers (even some of our personal heroes). Reactions from other designers (as well as a little praise) helped a lot. Reminded me of that super bad Kevin Costner film *Field of Dreams*: "build it and they will come".' 'But it always was personal. It was about us, and what we do, how we think,' adds Abbett.

DEVELOPMENT↑

(development)

TECHNIQUE { 'I think one person made all of these telephone box sex cards. We wanted to integrate this style for our "Contact" (Marketing) section. Work still in progress. Another idea: The new business section — we do anyone, anywhere, anytime...' says Abbett.

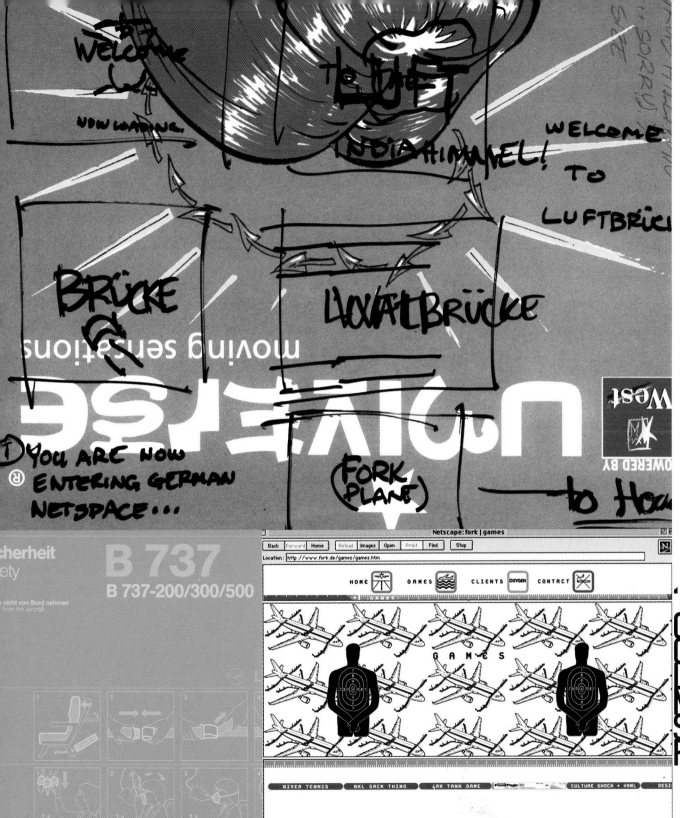

'Sketches are usually the notes that we make for planning sequences for user interaction or bookmarks for our ideas – we don't spend a lot of time making detailed sketches or roughs. Once we have a basic theme or graphic situation, we can usually describe it with some words or a quick sketch. Flyers are great sketch sheets… instant colour and spontaneous accidents with background typography!'

FORK uses sketches to organise the editorial content of the site. 'In developing the navigation, we scribbled the frame raster as well as subsequent combinations of framing to determine where the user will interact with menu options,' explains Linderman. A typical new project begins with text and picture sketches to describe the outside parameters and define a structure to work within. 'The good idea or metaphor usually comes unexpectedly, a lot of times at night, or when we're sitting together in a bar doing free association for a new kill game,' he adds.

'GERMANY WAS A GREAT CULTURAL CENTRE, BUT ITS CURRENT IDENTITY CRISIS AND LACK OF DIRECTION FOR THE NEXT MILLENNIUM IS A CONSTANT SOURCE OF INSPIRATION FOR US.'

'MUCH OF OUR INSPIRATION COMES FROM QUESTIONING THE SERIOUSNESS OF CURRENT MEDIA PROGRAMMING OR PROPAGANDA AND JUXTAPOSING THIS SERIOUSNESS WITH HUMOUR OR A HEALTHY LACK OF RESPECT. WE USE BIBLE PASSAGES INSTEAD OF LATIN FOR OUR DUMMY TEXT AND TURN RED INTO YELLOW.'

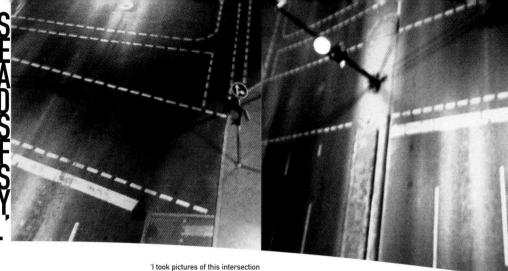

BERLIN INTERSECTION

'I took pictures of this intersection in Berlin because I was struck by the obviousness in which the different lines described the flow of traffic. I was also completely ripped at the time. But still, the lines are cool. Everything is so nice and orderly,' enthuses Linderman.

DEVELOPMENT

(development)

When it comes to building/designing and maintaining/adding new sections, Linderman describes the process as 'carefully planned chaos'. 'We usually all have totally different (but all awesome) ideas, and it's down to the survival of the fittest. If somebody can persuade (shout down) the others or manage their downtime better and create the first beta they win,' he laughs. But Abbett provides a dissenting voice: 'Sure if an idea that someone proposes is half-baked then we reconsider it, maybe try to add to it. But since all of us are quite different from one another, it's important that we all throw ideas off of each other. It's an evolution process.'

RECYCLING

'Everything we read and see is based on learned or remembered associations. By changing the context or modifying important elements of an icon we can change the meaning and recycle the sentiment or associations of the reader. In doing this, we create an emotional reaction. Most of these are personal snapshots that have been recycled in one way or another into the site,' says Linderman.

'We take pictures and video, sound bites and all other things that happen which strike us as being worthwhile and try to use them in our work. That way it's more personal, has more depth, and is not so generic, like... stock photography,' says Abbett.

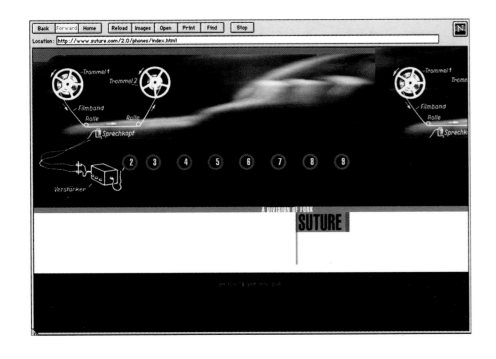

Both Linderman and Abbett find inspiration in everything: 'Our commercial work, the stuff we hear and see in clubs (we love music, especially electronica, breaks and drum 'n' bass) lots of analogue zines and art-sites as well as the web itself... grrl zines have great content ideas,' says Linderman, while Abbett puts it down to 'whatever seems to be pushing us that week, it could be having great weather or the latest media hype like the death of a celebrity. Inspiration comes from our immediate surroundings... with the communication technologies evolving, more events are exposed to us that have not much to do with our day-to-day living, therefore bringing more immediacy to distant concerns.'

'THE WEB IS NOT MASS MEDIA. EVERYBODY KNOWS THIS EXCEPT THOSE TRYING TO MAKE MONEY. IF ONE SETS OUT TO MAKE MONEY, THE SITE BECOMES GENERIC AND BORING.' TO MAKE IT WORTHWHILE, IT SHOULD BE PERSONAL. WHEN I LOOK AT A SITE, IT SHOULD BE ABOUT ME READING ABOUT YOU. EVEN ON A PURELY INFORMATIONAL LEVEL, I WANT DETAILS.

HARDWARE

> 'Jeremy told a paper recently that he believed in the Macintosh the same way that some people, say, believe in Scientology. We try and cultivate some smouldering conflict by maintaining some PCs too. Unfortunately they need operators that work on them,' says Linderman.

SOFTWARE

> 'Software is pretty old-school here: Director, FreeHand and Illustrator, Photoshop, Debabelizer, Premiere and Infini-D, BBEdit, Golive Cyberstudio and Baby Gif-builder.'

FORK UNSTABLE MEDIA
JULIUSSTRASSE 25, 22769, HAMBURG, BUNDERSREPUBLIK
TEL: +49 40 432 948 12 FAX: +49 40 432 948 11
E-MAIL: da5d@fork.de E-MAIL: abbett.j@on-line.de
CLIENTS INCLUDE: Beiersdorf AG, b&d Verlag; Spar
Deutschland; Hamburger Sparkasse Astra Brauerei;
Container Records, Hamburg; Schultz & Friends, N.A.S.A.,
Hamburg. [SEE PAGE 147]

Me Company, set up 13 years ago and housing 6–8 regular members, is one of the world's best-known exponents of computer art, its work instantly recognisable through its digital make-up and artistry. For the past five years all the studio's output has been digital and has attracted a wealth of clients – ranging from Nike and Diet Coke to Björk and obscure techno labels – on projects as small as flyers and labels, and as large as worldwide advertising and marketing campaigns for bands and products, consisting of packaging, billboard posters, digital films and videos.

While the company has its critics — people who claim its work is 'computer art' and has little value in design terms — Alistair Beattie, Me Company producer and co-director of the Makihara animation, refutes that body of thought: 'We regard our work first and foremost as design. We're really into having a good time, working with cool people, thinking intelligently and being original. Everything else doesn't matter. People's preconceptions and narrow opinions are annoying but there are always people out there who "get it", so why worry? As the saying goes: "The less room you give us, the more space we've got".'

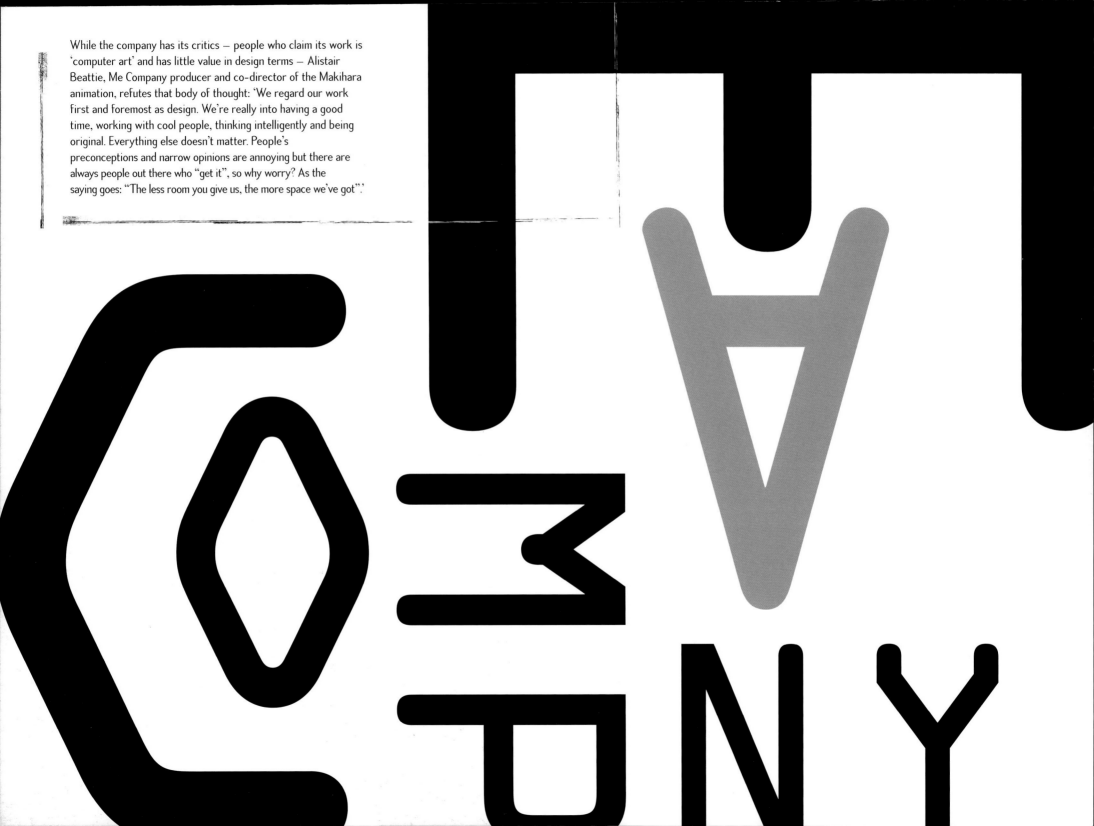

me company
electronic workshop

...PREPARE TO RECEIVE DATA...DIGITAL COWBOY ON LINE...PREPARE TO RECEIVE DATA.

CHARACTER DEVELOPMENT

Datababies are receiving satellites. They are the produce of the homestead, DataFarm. They're good robotic citizens of the Federation, harvesting raw materials of knowledge which they transmit back to the homestead.

HOWDY!

Let me introduce you to

a **DataBaby**.

He's our lifeline to

knowledge. He ac[]

satellite, gathe[]

knowledge from []

DataWorl[]

It's my job,
and I do the best I can.

THE VILLAINS

Dog of DogWorld is the villain in this story. He is a mean irascible character with only one purpose in life – to eradicate knowledge throughout the Universe by attacking the floating Datababies and the knowledge they harvest. He has an aggressive head with a massively armed mine for a body but he is confined to his kennel since losing the use of his body. Dog hates knowledge. It makes him miserable.

54

A·PLUS DATA FORCE

0
100
10100
0

01 39/a6e

CAP: 2000 TERABYTES

DigitalGirl Data

No.2 in a series

COWGIRL

Digital Cowgirl is a pink-skinned digital girl who wears her own customised DataFederation Programmers College uniform and gravity boots. She and her partner aren't 'rooting, tooting, shooting' cowboys but more like homesteaders. The DataFarm on which they live cultivates knowledge for everyone to benefit from.

'IT'S NOT ROCKET SCIENCE TO FIGURE OUT THAT A JAPANESE POP RECORD SHOULD BE CLEAN, COLOURFUL AND THE CHARACTERS SHOULD BE CUTE AND LOVEABLE.'

EARLY WORK

Sketches by Paul Garner of Digital Cowboy: a cheeky, happy character who is half man and half machine. From access panels in his head you can see the wires and transistors that are his brain.

(background) BACKGROUND

International record company WEA Japan came to the group with an open brief which asked for a Me Company treatment to an album by Japanese artist Noriuki Makihara. Virtually unknown in the West, Makihara is huge in Japan and Asia, having sold some 26 million albums. But for his first collection of songs in English, entitled 'Love Letters From the Digital Cowboy', he wanted to do something that was quintessentially English and differentiated this album cover from his Japanese output.

'We took the view that trying to sell a pop star to 16-year-olds is the same the world over, it has to have energy, vibrancy and a sense of fun,' says Beattie. 'No one had any preconceived ideas about what it might be about or look like,' he adds.

MAQUETTES

Clay fibre maquettes of all the characters by Petria Whelan were cyber-scanned: shown here are Digital Cowgirl, the female partner of Digital Cowboy, and Datababy. The Databables orbit DataWorld, the planet inhabited by Digital Cowboy and Cowgirl, and through the knowledge chip in their stomachs transmit data back to them.

(the project)
THE PROJECT

Digital Cowboy's pale blue hyper-teflon skin is pitted for special non-stick quality to protect him from the ravages of flying around the Micro-verse. Blue symbolises his ultra-micro maleness. He wears a cowboy hat made of wires and electronics and the uniform of the DataFederation Programmers College. The Scooter is a genetically engineered cross between a Venusian thwag-duck and Suzuki motorbike with jet intakes and rockets, whose head is a cross between a cow and a horse with the horns swept backwards to form handlebars.

'Taking the title as our inspiration we decided to create a Digital Cowboy, give him some supporting characters and design a narrative and a world to live in. So we had several characters, Digital Cowboy, Digital Cowgirl, The Datababies, The Dog of DogWorld and Scooter. The first thing we did was write a short story featuring the characters to give us all something to think about. It is really helpful when you're making decisions to have an overview about what you're trying to achieve. A narrative is like a spine running through the work,' explains Beattie. The short story centred around the adventures of Digital Cowboy, a space homesteader who lives on DataWorld and sports a modular body like a robot which is in the shape of an 'M'. His arms are tubular, and his boots are more like space boots than cowboy boots, with built-in grooves to lock him into position on his Scooter. 'The client was very exited by the written treatment and commissioned the artwork for the album, singles, promotion, advertising and point-of-sale campaign,' he adds.

From the original commissioning of the album cover, the project was to eventually expand to a huge campaign; applications of the characters on Tokyo metro tickets, seven-metre-high billboard ads and posters on the Tokyo subway, along with a 15-second ad and eventually the screening of the computer-generated video made from stills on jumbo screens in the city. 'The best thing about our client was their total openness to new ideas along the way. When they saw the digital characters coming to life, stage by stage, they immediately saw the potential in animating them. It meant that they could really follow through with the idea that these characters would completely represent the project in public,' says Beattie.

'IT IS AN INTERESTING PHENOMENON OF DIGITAL WORK THAT AS THE PROJECT TEAM GETS LARGER, LESS GETS DONE. WHEN WE STARTED THE WORK WE WERE DEALING WITH (FOR US) RELATIVELY UNKNOWN TECHNOLOGIES. WE LEARNT A LOT ALONG THE WAY.'

ver.1.0E · LOVE LETTER FROM THE DIGITAL COWBOY · Makihara

1100·1111·10110·101

1. Hey Yo!
2. Cowboy
3. The Lover In You
4. Running Out Of Daydreams
5. Day And Night
6. 9 x Forever
7. My Eyes Adored You"
8. I Can Feel Your Heart
9. Secret Heaven (Album Mix)
10. Limit's Of Love
11. I'm Not Gonna Fall In Love
12. Bye Bye Cowboy
13. If You Believe In Love

ver.1.0E

LOVE LETTER
FROM THE
DIGITAL
COWBOY

DATA

1100·1111·10110·10   0·10100·101·10010

1100·1111·10110·101·1100·101·10100·10100·101·10010

Noriyuki Makihara

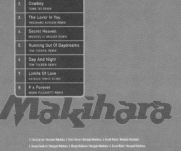

2. Cowboy
TOWA TEI REMIX
3. The Lover In You
YASUHARU KONISHI REMIX
4. Secret Heaven
MICHAEL H. BRAUER REMIX
5. Running Out Of Daydreams
TOM TUCKER REMIX
6. Day And Night
TOM TUCKER REMIX
7. Limits Of Love
SATOSHI TOMIIE REMIX
8. 9 x Forever
MARK PICCHIOTTI REMIX

Makihara

1. Chris Cannon / Noriyuki Makihara, 2. Towa Tei / Noriyuki Makihara, 3. Chris Cannon / Noriyuki Makihara
4. Randy Goodrum / Noriyuki Makihara, 5. Wendy Waldman / Noriyuki Makihara, 6. Chuck Kentis / Noriyuki Makihara
7. Rob Crane / Randy Nolan, 8. Linda Thompson / Noriyuki Makihara

NORIYUKI MAKIHARA · LOVE CALLS FROM THE DIGITAL COWGIRL

1. I'm Not Gonna Fall In Love
SATOSHI TOMIIE REMIX
2. Cowboy
TOWA TEI REMIX
3. The Lover In You
YASUHARU KONISHI REMIX
4. Secret Heaven
MICHAEL H. BRAUER REMIX
5. Running Out Of Daydreams
TOM TUCKER REMIX
6. Day And Night
TOM TUCKER REMIX
7. Limits Of Love
SATOSHI TOMIIE REMIX
8. 9 x Forever
MARK PICCHIOTTI REMIX

LOVE CALLS
FROM THE
DIGITAL COWGIRL

Makihara

Noriyuki Makihara

ᐯ
DEVELOPMENTᐯ
(development)

NARRATIVE

Me Company wanted to keep text on the album covers to a minimum as it would all be in English, so only necessary text such as track listings is on them, leaving the depth of the narrative to be discovered through the posters and video.

'After devising the narrative and its characters, Paul Garner, a storyboard and conceptual artist working for Me Company, sketched the characters, which were then transformed into clay-fibre models by Petria Whelan. These were the characters' first three-dimensional incarnation, defining the principle structures of the faces and heads. The character models were then cyberscanned to produce high-density poly meshes by Cybersite Europe. These were sent to 3D Technology and decimated into Macintosh file formats for Me Company. The details of the characters were then worked on, using Infini-D, Form-Z, Backburner and Photoshop, all in-house at Me Company,' recalls Beattie.

The resolution of the generated images ranged from 72 pixels square to 10,000 pixels wide. Me Company worked on the basic models, adding extra builds with the help of CG modeller Martin Gardiner. The final models were a composition of the cyberscanned material and builds lit and rendered in the various poses required for the project. This work was used for the print and poster campaigns.

For the four-and-a-half-minute video, the final computer-generated characters were transferred from Me Company's systems to animation company Lost In Space's systems by working closely with a team of eight animators. The resulting video includes wonderfully expressive character animation by freelancers Daniele Colajacomo, Jeff Bastedo and Mark Powers; fantastic flying sequences by Paul Simpson, and 'deeply funky' dance routines by Stuart Gordon. All the original characteristics for Me Company's characters were retained in the animated version, which took ten weeks to complete. 'It was a difficult process but eventually they managed to capture the look and feel of the print campaign,' says Beattie.

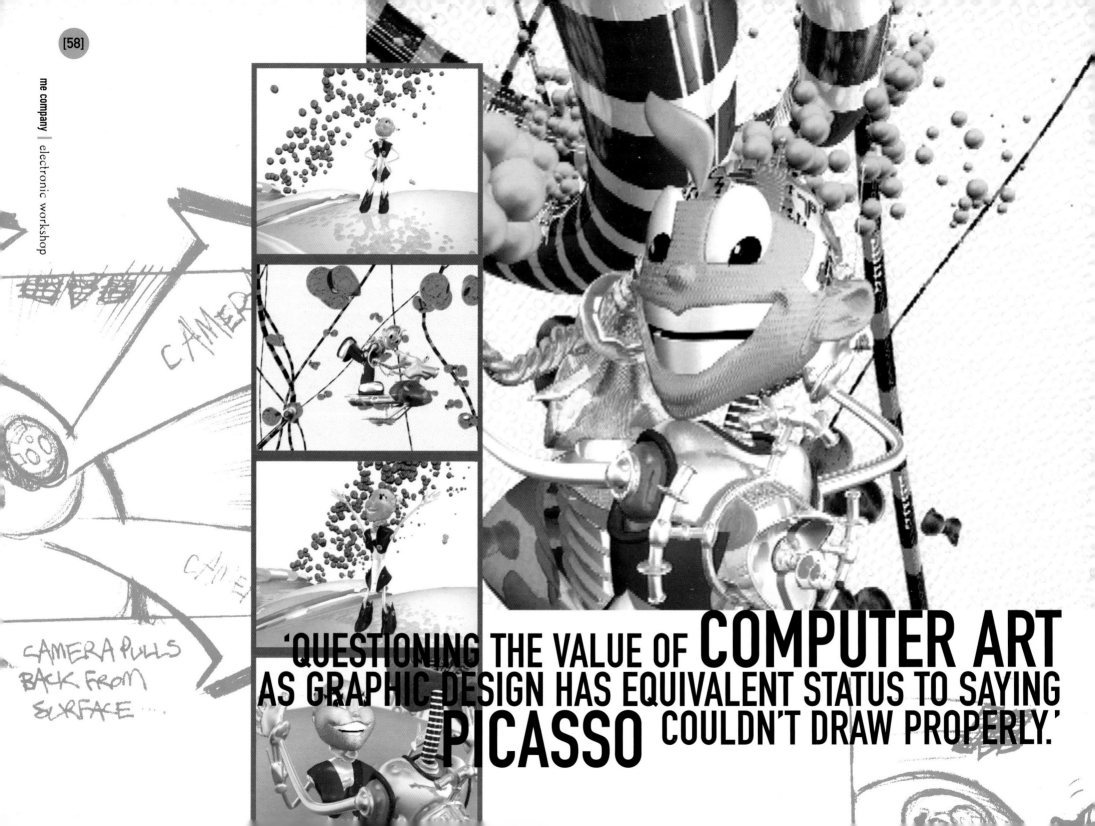

'QUESTIONING THE VALUE OF COMPUTER ART AS GRAPHIC DESIGN HAS EQUIVALENT STATUS TO SAYING PICASSO COULDN'T DRAW PROPERLY.'

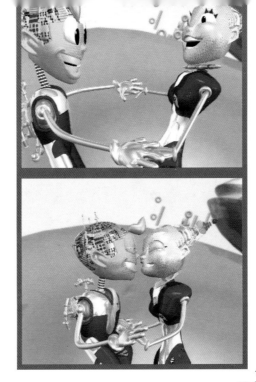

REVEAL FL...

(technique) and KIT ∨

'People get preoccupied with technique and software — but they don't think a carpenter's artistry lies in his tools so why should a designer's be seen as lying with the computer?' comments Beattie, adding that the 'development of a project at Me Company isn't really linear. While it all comes under the art direction of Paul White, co-founder of Me Company, any number of people, inside and outside the company, will be working on different elements concurrently — we don't have rigid stages, it's much more organic than that.' When pressed, he gives a broad outline of techniques and development: 'Text (narrative), character design (drawing, redrawing, re-redrawing), sculpture (generating maquettes from illustrations), scanning (digitising the maquettes), cleaning and decimating (making the scans function inside our CG systems), surfacing (giving the models surface attributes), light rendering, retouching, graphic design, layouts, artwork colour balancing, conversions and print tests.'

ANIMATION

'All the characters for the print project were rendered and generated in three-dimensional models, a process which lends itself to animation. The resulting computer-generated film was very fast and frenetic and worked very well in Japan.' says Beattie.

HARDWARE

> MACINTOSH FOR THE PRINT CAMPAIGN

> SGI FOR ANIMATION

SOFTWARE

> INFINI-D

> FORM-Z

> BACKBURNER

> PHOTOSHOP

> FREEHAND

> PRISMS

Me COMPANY
14 APOLLO STUDIOS, CHARLTON KINGS ROAD
LONDON NW5 2SA, UK
TEL: +44 (0)171 482 4262 FAX: +44 (0)171 284 0402
E-MAIL: meco@meco.demon.co.uk

CLIENTS INCLUDE: Nike; Coca-Cola; Warner Records; Polygram
Records. [SEE PAGE 151]

JOHN J. HILL, an alumni of the Michigan School of Art, is based in New York where he started out as a freelance, working on artwork for CD-ROM titles, CD packaging, websites and comic books. In 1996 he set up design studio 52mm with Marilyn Devedjiev and Ron Croudy. They work on many different forms of media – including film and video, print, web design, CD-ROM and music: 'My main interest is the conversion of traditional arts from their original forms to the digital medium – music, illustration, photography, painting, film... the list keeps growing,' he explains.

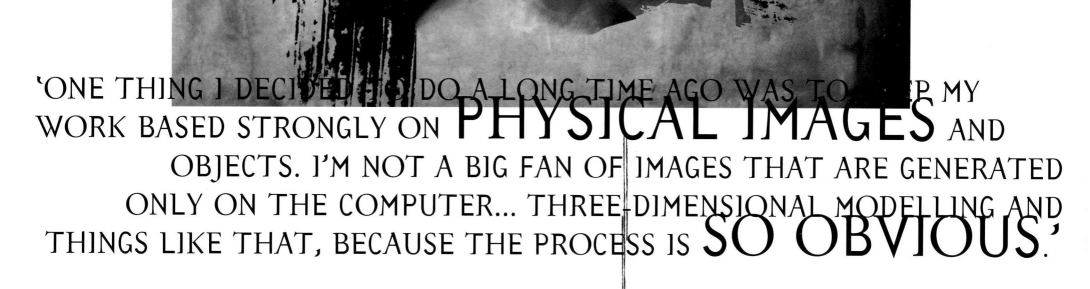

'ONE THING I DECIDED TO DO A LONG TIME AGO WAS TO KEEP MY WORK BASED STRONGLY ON PHYSICAL IMAGES AND OBJECTS. I'M NOT A BIG FAN OF IMAGES THAT ARE GENERATED ONLY ON THE COMPUTER... THREE-DIMENSIONAL MODELLING AND THINGS LIKE THAT, BECAUSE THE PROCESS IS SO OBVIOUS.'

(background)

BACKGROUND >

Hill arrived at the *Ascension* project shown here through a four-month stint working on a website for Top Cow Productions, part of Image Comics. 'When I first took on the website project I was familiar with their different comics and characters, but over the course of doing the site I learnt a lot about every aspect of the company. Top Cow is well known in the comics industry as producing very high-quality work. After a few months of dealing with programming, image compression, uploading files and heavy design work, I needed to get some pent-up illustration energies out,' says Hill. 'I felt like doing a piece of art based on one of their properties,' he adds.

Image Comics was just about to bring out *Ascension*, a new comic about the conflict between two warring races, the scientific Mineans and the more aggressive Dayaks, who dwell in a kind of extra-dimensional realm which has a symbiotic relationship with our world. A complex plotline that would make a great 'X File' involves the Chernobyl disaster, missing knowledge found in the scattered pages of an apocalyptic book and their relationship with American scientist Lucien, who grows into a supernatural being with a difference. He becomes physically, emotionally and mentally affected by the book. Lucien and Chernobyl were to form the basis of Hill's illustration project.

COMICS

Cover images of the *Ascension* comic from Top Cow Productions, by David Finch, penciller/co-writer; Batt (Matthew Banning) inker/co-writer; and Liquid Graphics (Christian Lichtner and Aron Lusen) colourists. *Ascension* is copyright and TM 1997, 1998 Matthew Banning and David Finch. Published by Top Cow Productions and Image Comics.
The winged character is American scientist Lucien, the focus of Hill's piece.

(the project)

THE PROJECT

Hill hoped to persuade Top Cow to commission a piece which could be used as a pin-up, poster or trading card, but when his timing proved wrong he decided to go ahead and do *Ascension* regardless. He finds that there's little difference between doing a commissioned piece and one that he's doing on his own: 'A commissioned piece may have a few elements which the client requires to be in there, but other than that my process is the same,' he says. 'Originally, I started the piece based on *The Darkness*, one of Image's other properties. As I was working on the background textures I realised that this was shaping into something different than I planned, and the focus changed,' explains Hill.

TEXTURES

Most of these textures ended up as the black shapes in the final image. By varying levels and using Photoshop's dodge and burn tools Hill was able to make the contrast on them as close to black/white as possible.

'I'VE HAD A THING FOR ANGELIC FIGURES AND WINGS FOR A NUMBER OF YEARS NOW. I ACTUALLY HAVE A TATTOO OF AN ANGEL ON MY FOREARM. EVERY FEW YEARS ANGELS WIND UP MAKING THEIR WAY INTO MY WORK... I THINK THIS IS ONE OF THOSE OCCASIONS.'

As the main focus is the relationship between Lucien and the pages of the book that turn him into an other-worldly being, one of the things Hill wanted to do was give the piece a mythological, classical story feel: 'The book element helped me to accomplish this in a big way... the figure is both layered on top of the book and is also the book's "cover" at the same time. So, you could say that this winged figure (which is a cornerstone of classical art/storytelling in itself) is intertwined with this book and these pages, or is part of the story contained in the book,' he explains. 'There's a tension between the two — the black jagged areas, whether or not the figure is part of the book, the soft wings versus sharp edges and so on — but they're obviously closely bound,' he continues. 'I used colours that are a little "off", along with rough static textures as my version of radiation. There's a lot of little subtle things here and there that relate to the story.'

WINGS — The search for wings took Hill to the Museum of Natural History where he shot some of the stuffed birds.

(development)

As his work consists mostly of combining photographs, found images and objects to form the individual elements which will make up the final image, Hill began by exploring the main focus of the piece, and the types of images or elements necessary to achieve that focus: 'Sometimes some very specific items may come to mind from the start, while other times I have a very general feel in mind that I look for in my image elements,' he says. 'I enjoy putting different elements and hints into a piece, and really let the viewer take it from there. There are a lot of pieces of the *Ascension* comic storyline in here which hardcore readers will probably pick up on, but those unfamiliar with the comic may piece together a completely different story, using the same elements,' he says.

'I mostly go for quantity over quality at this phase. I'd rather have a ton of images to choose from and spend a minimal amount of time getting them than labouring for hours over a single photograph. There's plenty of time to get things to look just right later on in the process.' Having edited the images down to as few as possible before deciding on their placement, Hill then focused on putting the background together, before moving to the main focus of the piece. Secondary elements and final additions rounded out the piece.

'WHEN ILLUSTRATORS START TO USE A COMPUTER, THEY NEED TO UNDERSTAND IT'S NOT A MEDIUM UNTO ITSELF. IT'S NOT SOMETHING THAT WILL DO THE ART FOR YOU. IT'S A TOOL. IT'S A FANCY BRUSH. IT'S A NEW KIND OF PAPER. WHEN I SEE AN ILLUSTRATION THAT COULD HAVE BEEN DONE THROUGH "TRADI-TIONAL" MEANS, OR EVEN MAKES YOU THINK "HOW DID THEY DO THAT?" THAT'S A SUCCESSFUL COMPUTER-GENERATED IMAGE.'

The book is several layers of the same image, one in multiply and one in overlay mode. The multiplied layer lays down the basic shape of the book image, but in several areas it becomes too dark, so the overlaid layer brightens up some of the book's areas (especially the type and the texture on the spine), which were virtually untouched. Hill then brought the rest of the book down to around 30 per cent opacity, 'which gives that whole area a slightly lighter hue and helps it to pop from the background,' he says.

BOOK

(technique) and KITv

'A lot of my work ends up being trial and error, and the actual working process is important. I use the image elements as my "paints", which I move around on the canvas and mix with others to create new "colours". The image will come together on its own a lot of the time, while I'm moving the various elements around and changing their relationships to each other,' says Hill.

'One concept I use a lot is in taking the lines and contours of several images and blending them together to create the lines and contours of a new image. These "real world" objects are only small pieces of the puzzle I'm putting together, which ends up being a completely new visual.'

The *Ascension* piece is 8 x 11 at 350dpi, so three or four extra layers really start to slow down the process. The other reason is that it makes it too easy to go back and change things... I could tweak an image for months and not be completely satisfied. Flattening after a while helps to solidify things in my head, and forces me to move on to other areas.'

ASCENSION

'A lot of Lucien's physical attributes, such as the wings and fire, make it into the piece and I also tried to convey his anger with the figure's pose. The Chernobyl accident and the book pages wind up being the background and elements that tie the piece together... as in the story.'

HARDWARE

> POWER MAC 7100/66, 72MB RAM, 1GB HARD DRIVE
> POWER MAC 8600/300, 128MB RAM, 4GB HARD DRIVE
> ZIP DRIVE, JAZ DRIVE, MICROTEK HR11 SCANNER, WACOM TABLET
> CANON Z70W CAMERA

SOFTWARE

> ADOBE PHOTOSHOP
> SCANWIZARD
> A SNOOPY pencil and laserwriter paper

JOHN J. HILL
12 JOHN STREET, SUITE 10, NEW YORK, NY 10038, USA
TEL: +1 212 766 8035
E-MAIL: jinn@pop.inch.com
WEBSITE: http://www.inch.com/~jinn

CLIENTS INCLUDE: Acclaim Entertainment; Agency.com; American Express; Capitol Records; Virgin Entertainment; Flatline Comics; LaQuinta Filmworks; Axis Comics. [SEE PAGE 149]

exemplar número:

Catalan type designers Joan Carles P. Casasín and Andreu Balius run Barcelona-based Type foundry and design company **TYPEWARE®** in an unconventional and anarchic way. 35-year-old Balius and 29-year-old Casasín first got to play with new technology at the the CIEJ (Initiatives for Experimentation for Young People Centre), which enabled them to use Macintoshes and early versions of programs such as PageMaker and FreeHand. In 1993 the pair met and formed the (very little) design company of two: Typerware®, which practices graphic design and mountain bikes, say the duo.

'We started designing typography for fun and making some flyers and posters for the music underground scene for friends (no money), along with some CDs... a lot of work for friends — you know, freedom but no money, but it's cool... Our early works in typography, most of them published in the Garcia Fonts & Co. library, made us something of a name in the typographic scene in BCN and we get some interesting type-related projects in Barcelona — which is nearly impossible!' explains Casasín.

It's easy to see why they're doing so well. Their work is fresh, radical and eclectic, while remaining functional, intelligent and well-respected; for example, their Universitas Salamantini typeface, designed for the Salamanca University last year, won a Macromedia award for best type design. And while they may come across as whacky and decidedly non-corporate, there is no doubting the pair's seriousness and intensity about that project: 'We have taken an historical-archaeological approach to make a modern typeface working on the basis of a Roman-humanistic shape; it had nothing to do with nostalgia or revivalism but was based on the lettering reproduced on the walls of the University. Dating back to the 16th century, the history of these walls is amazing,' they say of the work.

As with many type designers, Typerware® do much more than create typefaces, expressing themselves and promoting their fonts through pamphlets, posters, and literature for themselves but also for clients as diverse as Editorial Anagrama, Accidents Polipoètics, Sol Picó Dance company, Museu Nacional d'Història de Catalunya, Town Halls (Sabadell, Mollet del Vallés, Montcada i Reixac), Fundació La Caixa, Fundació Miró, Institut Universitari de l'Audiovisual de la Universitat Pompeu Fabra and internationally-acclaimed Spanish performance group, La Fura dels Baus.

(background)
BACKGROUND

Originally commissioned to design a graphic image system for the La Fura show
entitled 'Manes', Typerware® were also asked to develop some kind of object
which would complement and support the show in some way. 'We eventually
decided to translate it in a book. We had dollar restrictions, so it was kind of
low-budget design. We looked for economy in the resources in order to
get the best for the least possible money,' say Casasín and Balius.

While La Fura translated the whole idea of the 'Manes' show, Typerware®
were given the space to propose any ideas they wanted: 'La Fura were
very open-minded and they never tied our hands and ideas up, it was a
dialogue with good feeling,' says Balius.

'MY ONLY RELATIONSHIP
WITH GRAPHIC DESIGN AND
TYPOGRAPHY BEFORE
DESIGN SCHOOL WAS
DRAWING THE NAMES AND
LOGOTYPES OF ROCK
BANDS I LIKED ON MY
SCHOOL TABLE.'

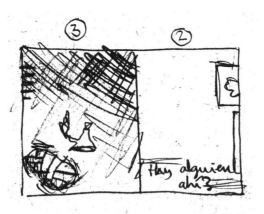

The combination of the FaxFont type with stills from the show and other graphic devices results in a book which acts as a piece of art in its own right, while echoing perfectly the frenetic, chaotic beauty of La Fura dels Baus.

(the project)

THE PROJECT

Unlike a concert programme or souvenir brochure, the purpose of the La Fura book 'is the translation of the concept of the show into an object. It is something like buying a souvenir of the show, but more than that it contains the essence and concept of the show. The idea was to devise something which contains its own art and offers the spectacle of "Manes" through the pages of a book,' explains Casasín.

'We hadn't any previous relationship with La Fura. It was commissioned directly by Pera Tantiñá and Elena Blanco (the creator and the production and communication manager). And they (together with Rafel Vives, performer director) were the people who talked with us, and said okay to our work,' he continues.

For their inspiration Typeware® looked to La Fura as a company; their experimentation and originality, but 'mainly the performance, the ambient and the sensations you get in "Manes", says Balius. 'You're there, taking part in the show... up and down, left and right, close to the performer feeling those sensations,' he enthuses.

DEVELOPMENT
(development)

The book's structure and elements of text, design and photography were developed in conjunction with the show: thus as the show progressed photos were taken which began to form the backbone of the book, but, as Balius explains, 'Each time we got new photos they were better than the previous ones. Even when we had all the book mounted we were still getting new photos and we decided to change some pages (more work!) at that very late stage.'

La Fura commissioned copy by Accidents Polipoètics (Rafa Metlikovec and Xavier Theros). This text was translated into Catalan, Spanish and English, and Typerware® highlighted sections of it 'for design and expressive content. There's no chronological line, so the text is without order, it's like the show: a zapping of emotions and sensations. It all happens simultaneously,' says Balius.

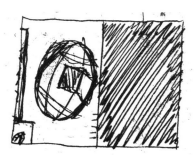

'Photography for the book was by Gol, who was chosen by Typerware® and the three members of La Fura who had commissioned them: Tantiñá, Vives and Blanco. We worked closely with Gol, discussing impressions about the show and how we thought the photos should be, but always respecting his decisions,' says Casasín.

WHAT WE WERE LOOKING FOR IN THE FAXED IMAGE WAS DRAMA AND EXPRESSION, AND IN THE END IT WAS THE PLASTIC IMAGE WE DECIDED TO USE.'

SKETCHES Typerware® began by doing thumbnails of the book, which although basic, show firm ideas and a distinct direction which they remained true to throughout the project.

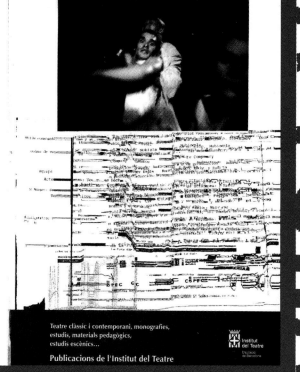

Teatre clàssic i contemporani, monografies, estudis, materials pedagògics, estudis escènics…

Publicacions de l'Institut del Teatre

Institut del Teatre
Diputació de Barcelona

LA FURA

PHOTOGRAPHY

Photography by Gol was chosen by Typerware® and members of la Fura. 'We worked closely with the photographer, changing impressions about the show and how we thought the photos ought to be, but always respecting his decisions,' say Typerware®. Duotones, used on most of the images, were built using Photoshop.

BUDGET

While the limited budget on the project meant Typerware® could not use full-colour on the book, they made great use of spot varnish and features such as two different blacks. Graphic devices such as the blue 'running-out' strip of a fax roll were also incorporated.

With photography and text in place, Typerware® and la Fura moved on to typography. 'This was obviously an important issue in the book, and we realised it had to express the frenetic drama of the show. We decided to design the FaxFont family. We had used some faxed type in a poster we'd designed around the same time and we thought it was a good starting point, like a sketch or blueprint for the La Fura work. So we took some characters from there and did more. The process is very easy (it was EasyType series!) — print type, fax and distort a lot, select the best, scan and trace and put it in the appropriate software — in this case Photoshop, Illustrator (or FreeHand) and Fontographer 4.1 and 3.5.1.'

The combination of this distorted and disorientating type with stills from the show and other graphic devices results in a book which acts as a piece of art in its own right, and echoes perfectly the chaotic beauty of La Fura. 'People who have attended one of their shows know exactly what the book is doing. The best thing that has been said about it is: "Yes, you can breathe La Fura's way of performing in those pages",' say Balius and Casasín proudly.

COVERS

The book's front cover has a story which continues in the back of the 'Guardes', or flyleaves. This story is essentially that of Pedro Manes, the 'Manes' of the show's title. Aside from pictures of Pera Tantiñá speaking with the actors in the early days of the development of the show and two bloody chickens, all the pictures are from the actual show itself.

Fura dels Baus   Manes

TYPOGRAPHY

'The typography was to be an important issue in the book. It had to express the drama of the show,' says Casasín. He and Balius built from characters which they had designed for a poster, arriving at the FaxFont family. 'The process is very easy (it was EasyType series!) – print type, fax and distort a lot, select the best, scan and trace and put it in the appropriate software, Fontographer,' they explain.

abcdefghijklmnopqrstuvwxyz

HARDWARE

> Apple Power Macintosh 7100/80

> Apple Power Macintosh 7500/100

> AGFA Studioscan

> Personal Laserwriter NT300

SOFTWARE

> MM FreeHand 5.5

> Adobe Photoshop 3.0

> Fontographer 3.5 and 4.0

ANDREU BALIUS AND JOAN CARLES P. CASASÍN

TYPERWARE®, ANSELM CLAVÉ, 11, 08106-SANTA MARIA DE MARTORELLES, BARCELONA, SPAIN TEL/FAX: +34 93 579 09 59

E-MAIL: typerware@seker.es

WEBSITE: http://bbs.seker.es/~joanca

CLIENTS INCLUDE: Fundació Miró; La Fura dels Baus; Editorial Anagrama; Accidents Polipoétics; Fundació La Caixa; Sol Picó Dance Company; Museu Nacional d'Història de Catalunya. [SEE PAGE 157]

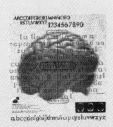

Jason

24-year-old **Jason Statts** only left Savannah College of Art and Design in 1996, but he already has a number of awards for his work – including two pieces in the prestigious Society of Illustrators Annual Student Scholarship Competition, 1997, in New York City and an entry in the 1997 Society of Illustrators of Los Angeles Annual Student Competition.

When you bear in mind that these two metropolitan areas probably house most of America's best illustrators, you realise how important those achievements are. Statts is beginning to build up a client list which includes *The Village Voice*, *Internet Underground* magazine, *Musician* magazine and N Soul Records, but when he was asked to contribute to this book he decided he'd like to originate a piece especially.

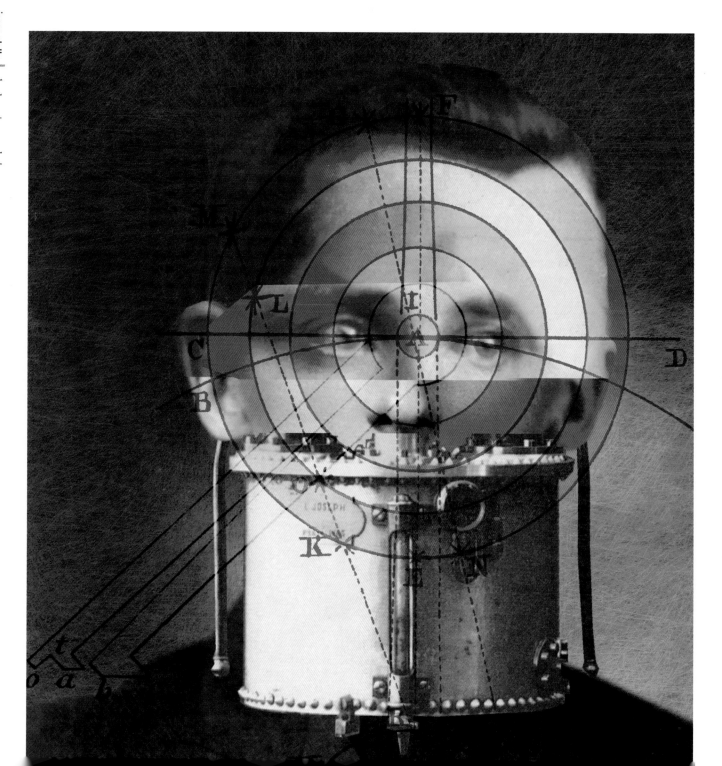

(background)
BACKGROUND˅

When Statts began to think about a project for the book, he came back to a subject he has frequently thought about over a number of years: 'The human psyche, what makes us tick, what defines normal and abnormal, and who ultimately decides what is right and what is wrong. It has always amazed me how fine the line is between sanity and insanity. So the basic idea of the piece is targeting mental disorders,' he explains.

'Most of the influences for this piece came from outside sources — television, magazines and so on. We're constantly being bombarded with information on mental illness and hear horror stories of people going crazy and walking into the nearest fast food joint or post office to open fire on men, women and children. This piece was a response to all of that. Just my way of dealing with it, responding to it, I guess,' says Statts.

SKETCHES

From the beginning, Statts had a very clear idea of what he hoped to accomplish with the final image, which is why these initial sketches bear a strong resemblance to the completed project.

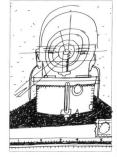

PAT.-PENDING
MADE IN U. S. A.

1   2   3   4

MAY 20 96
JUL 3 96
OCT 21 96
FEB 25 97

'I THINK ALL THESE ELEMENTS COMBINED LOOK LIKE AN IMAGE ABOUT THE HUMAN CONDITION, WHETHER IT BE AN ACTUAL ILLNESS OR JUST A HEADACHE. IN THE CONTEXT OF A STORY OR ARTICLE I THINK IT WOULD SERVE ITS PURPOSE WELL AND BE EASILY UNDERSTOOD.'

(the project)

THE PROJECT >

SOURCE MATERIAL

Statts wanted the object over the figure's mouth to impart a scientific feel – 'because science is involved in both studying brain chemicals and in diagnosing mental disorder' – and turned to this image of part of a helicopter motor. The dates, taken from an old library check-out card, were added to the bottom right-hand corner of the image to symbolise time and memory.

From the outset, what Statts was trying to represent was the intricacy and fragility of the human mind and psyche: 'There is so much that goes into making us individuals, and if one microscopic detail gets out of whack we could very well snap,' he comments. What came to mind immediately was the idea of the bull's-eye or target on someone's head, which he began to sketch, thinking about the elements he would use as he did so.

In its structure and composition, this project was very typical of Statts' working process. 'I usually don't create any elaborate compositions, they are very straightforward. That's not to say there isn't a lot going on in the piece, just that the overall compositioning of it will remain relatively simple,' he explains.

jason statts | electronic workshop

PORTRAITS

Statts uses as many as five individual pictures when making up a figure or portrait. Here two, taken from a royalty-free CD-ROM collection by Toronto-based company Eclecticollections, were used because 'I knew I'd be covering quite a large portion of the facial area with another object', says Statts.

DEVELOPMENT

(development)

Building from a background created of scratched and re-scratched colour layers in Photoshop, Statts began to assemble his collage. 'Since I work in an assemblage/collage style, it is hard for me to pinpoint exactly what the final image will look like. Most of the time, I don't know what I'm going to use in the final piece until I actually use it, so there is always an element of surprise – both for me and whoever has commissioned me!' he says.

Statts started with two portraits and images from a royalty-free CD-ROM by Canadian image library Eclecticollections, and began the search for more abstract elements: 'I have a pretty extensive collection of "junk" that I use as source imagery — everything from stamps to doorknobs, paper, paint chips, bingo game pages, fuse boxes and lots of antique-looking things. These come from taking things from the streets, and if I can't take something I'll photograph it.'

'THE COMPUTER ALLOWS US TO COMBINE PAINTING, PHOTOGRAPHY, COLLAGE, MONTAGE AND JUST PLAIN TRASH IN OUR IMAGES AND WHAT'S COMING OUT IS BEAUTIFUL. PEOPLE ARE REALLY GOING NUTS AND TRYING OUT ABSOLUTELY INSANE TECHNIQUES – AND GETTING AMAZING RESULTS.'

T. PENDING
DE IN U.S.A.
1   2   3

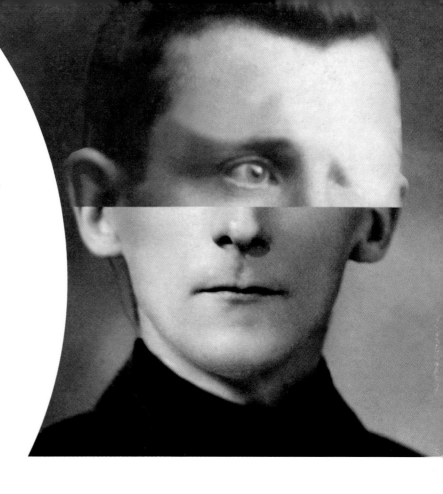

'IN TERMS OF ILLUSTRATION, THERE'S NOTHING MORE FRIGHTENING THAN THINKING YOU HAVE JUST LOST A CLIENT THROUGH LACK OF COMMUNICATION.'

(development)

A diagram of the solar system formed the base for the bull's-eye, part of a helicopter motor placed over the mouth was used to suggest scientific process, pages from an old encyclopaedia and numbers from a library card deal with knowledge, time and memory, and the ruler — an old tailor's gauge Statts found in a thrift shop and has used in many of his illustrations — 'was used to symbolise how scientists measure our intelligence with tests and refer to numbers and statistics to gauge whether we are inside or outside the norm,' explains Statts. And the 'S'? That just stands for Statts.

TECHNIQUE | 'One of the major things I always seem to do is switch out the eyes — it's amazing what a pair of eyes can do — it always seems to add so much to the piece,' says Statts.

PROCESS⟨ A diagram of the solar system
provided Statts with the bull's-eye
he was looking for to represent
the 'targeting' involved in
diagnosing mental disorder.

(technique) and KIT ᵥ

As Statts likes to 'play up the collage look and play
down the computer look,' he will scan in background
'textures' created through traditional painting, and use
'I will usually have at least three "colour" layers in any one image, these as a basis for the overall image. 'I use a lot of lay-
and will "scratch" these layers with the eraser tool and brush tool ers. I keep just about every little piece of the image on
to create different textures and colours. I do quite a bit of an individual layer, so that I can fully explore the possi-
"scratching" on a lot of layers and elements, resulting in an inter- bilities of where things could go, and how the different
esting texture. I use the layering effects (in the layers palette, ie: layering effects work the best for each element,' he
multiply, overlay, screen, and so on) on just about everything in explains.
the image, or at least try them out. I also use brightness/contrast
and levels on just about every project. I really don't get very
technical, resulting in work which, for the most part, looks like
plain, old-fashioned collage, like a computer wasn't used at all,
and I like that,' says Statts, adding: 'the computer just makes my
techniques easier and faster.'

TECHNIQUE⟨ The muted, earthy tones of the
project are typical of Statts' work:
'I like to illustrate dark subjects, I
find them more fun, and as happy
colours are inappropriate to those
subjects I don't use them unless
the client asks outright for them,'
he explains.

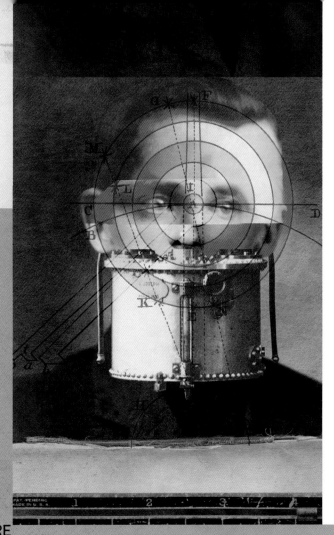

'I KNOW MORE ABOUT PHOTOSHOP THAN THE AVERAGE JOE ON THE STREET, BUT I DON'T THINK I DO ANYTHING TOO SPECTACULAR WITH THE PROGRAM, INSOFAR AS TECHNICAL ISSUES ARE CONCERNED. I JUST TRY DIFFERENT THINGS UNTIL I GET THE RESULT I AM LOOKING FOR, OR A CLOSE APPROXIMATION. IF IT FEELS RIGHT AND LOOKS RIGHT, IT MUST BE RIGHT. **RIGHT?** RIGHT.'

REPRESENTATION

Pages from a 1930s encyclopaedia were used to 'represent knowledge: I love aged books, paper and photos – there's nothing quite like the appearance of the yellowed pages', enthuses Statts.

HARDWARE

> POWERMACINTOSH 8500/180 WITH 72MB RAM

> UMAX POWERLOOK II FLATBED SCANNER

SOFTWARE

> ADOBE PHOTOSHOP 4.0

JASON STATTS
POST OFFICE BOX 16626, SAVANNAH
GEORGIA 31416, USA
TEL: +1 912 355 8398 FAX: +1 912 238 2459
E-MAIL: thinearth@worldnet.att.net

CLIENTS INCLUDE: *The Village Voice*; *Internet Underground* magazine;
The UTNE Reader; *Musician* magazine; *Jazziz* magazine; N Soul Records.
[SEE PAGE 148]

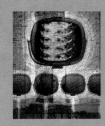

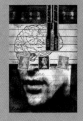

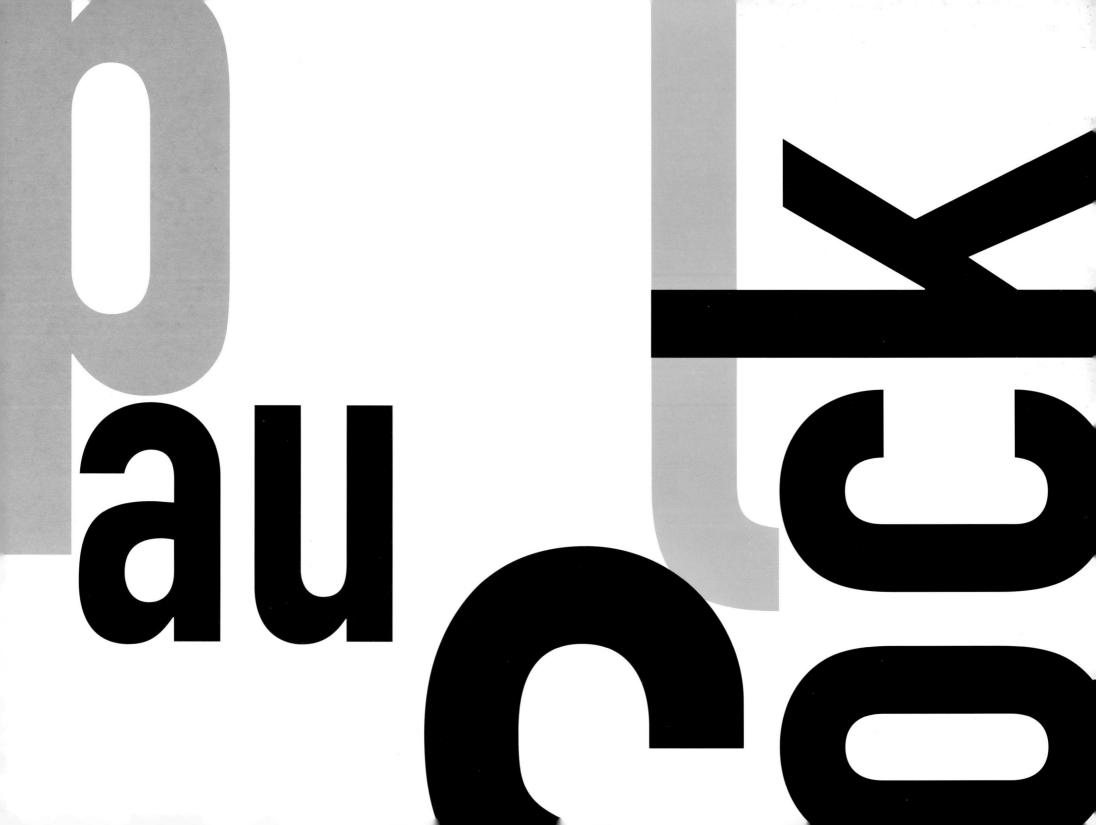

London-based illustrator **Paul Shorrock** has been working as a professional illustrator for 14 years and started to produce his work digitally in 1995. While the computer has not changed his style greatly, as he effectively builds up collages using different media, it has hugely facilitated his work and enabled him to overcome physical limitations such as colour registration, personal control and changes to the work.

'In the past my technique allowed fairly limited colour, because I'd basically be tied to using whatever colour paper I could find to photocopy on to, whereas now I have a choice of millions of colours. I also have a lot more control over my work, it fits in seamlessly with the client's equipment and the best thing is there's no physical piece of artwork, which takes away the tension of working on a piece that could so easily be damaged,' he comments. 'Another obvious advantage with the Mac is you can save different versions of a project, keep work in progress and make changes very easily, which in the past was impossible without taking the whole thing apart and virtually starting over,' he adds.

'THE FACT THAT IT WAS A SERIES MEANT THAT I HAD TO **PACE MYSELF**, NOT USE ALL THE BEST IMAGES IN ONE GO, AND I HAD TO BE AWARE OF NOT PAINTING MYSELF INTO A CORNER WITH TECHNIQUES AND COLOUR, SO THAT IT AFFECTED THE **OVERALL DESIGN** OF THE PROJECT.'

Last year, *The Sunday Times Magazine* commissioned Shorrock to illustrate a series called 1000 Makers of Music; a project consisting of six magazine-sized supplements and a binder building up into an A-Z encyclopaedia about musicians, composers, instrument makers, critics, producers and so on. The magazine needed six cover images plus five or six smaller images for the inside pages where a portrait would be inappropriate or unavailable; for example categories such as Anonymous, The Symphony Orchestra and Nightclubbing.

(background)
BACKGROUND

'The unusual thing about this project was having to use photography, which I rarely do. *The Sunday Times* supplied a load of photos and I decided which to use. I went for iconic images, as well as people whose music I like, always thinking about the balance over all six covers, covering classical, pop, jazz, country and so on. I had quite an open brief, they gave me the size and photos and wanted me to use quite bright dynamic colours, but other than that I had a free hand,' says Shorrock.

ROUGHS

Having decided on images for the S-Z category, Shorrock then began to lay out rough black and white scans of them: 'At this stage I'm just making sure that the images are relevant and work well together,' he says.

The large scale of the project dictated a style which would span the whole series of six covers and a number of inside illustrations, so Shorrock decided on a combination of various elements such as drawn images and photographs, all placed on a musical manuscript, which would be the background for all six covers. 'That became the formula for the whole series, and the brightly coloured manuscript background helped tie it all together, giving all the elements something to sit on. It's a score I found in a second-hand shop coloured in two colours to give it a slightly three-dimensional effect. Then I just play around with the images, making sure they sit together in a harmonious way, getting the shapes right, not like a jigsaw but so that one shape plays off another, allowing the eye to follow through various elements. So all the images are scanned in as separate elements to allow that movement and layering.' For the entire project, Shorrock made optimum use of Photoshop 4.0, and indeed used no other software in its production.

(the project) THE PROJECT

The black and white photos on the project were supplied in four-colour then the colour balance adjusted to achieve different tint effects. Photoshop filters were used sparingly to improve the photographs: 'For example, on the Velvet Underground picture, which was a bit flat and fuzzy, I used the unsharp mask filter which heightens the contrast between adjacent pixels, giving it a grainy quality with better focus and contrast,' explains Shorrock.

TECHNIQUE

'IF YOU'RE SET **PARAMETERS** THEN YOU HAVE TO WORK WITHIN THEM, BECAUSE IF YOU'RE GIVEN FREE REIN AND EXPRESSION THEN YOU'RE A FINE ARTIST, WHEREAS ILLUSTRATION'S **CONSTRAINTS** MEAN THAT YOU HAVE TO SOLVE A PROBLEM AND COMMUNICATE A SPECIFIC IDEA.'

Having decided on the structure, Shorrock had to ensure all the elements would look sufficiently different and stand out from each other. This was done with background colours, positioning and manipulation of the individual elements. 'Aside from making each element correspond to its particular section of the alphabet, I also had to be aware of properly representing women, ethnic minorities, and all the genres from rap to folk and classical. Musical instruments are so visually interesting and diverse that I decided early on to use those, and looking at them to me suggests sound and music, so I think they give the piece a dynamism,' he says. An added sense of movement was incorporated through such elements as dancers and musicians.

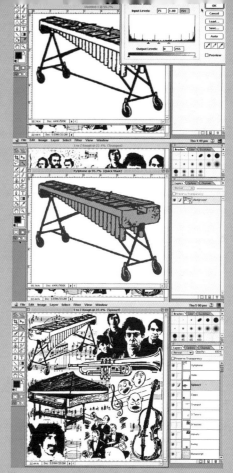

'The contemporary photos had to go in and they contrast well with the historical collage images, which don't necessarily suggest up-to-date, of the minute ideas. With images taken from old books there's a lot of cleaning up and sharpening of lines to be done, and even on the photographs, like the Velvet Underground one, you can acquire an added depth by doing an almost imperceptible blurred opaque shadow behind it. And white haloes are also useful for making things stand out from a dark background,' continues Shorrock.

(development)

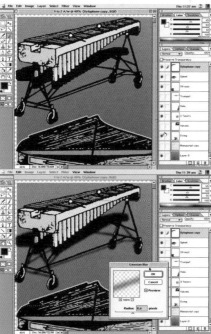

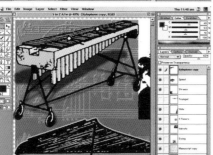

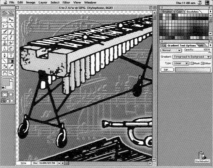

TECHNIQUE {

There are many layers and images making up the whole, so it was important to get as much depth as possible into them. One way of doing this is through the creation of drop shadows, as these pictures show. Shorrock explains: 'With the xylophone, to get the effect of its standing I added a transparent shadow on the layer below it. The shadow cast on the "floor" had to be redrawn and distorted to fit the horizon. I used the Free Transform tool to manipulate the shadow, which was created by copying the xylophone layer and filling with black. Once in place it was given a Gaussian Blur filter to soften the edges and the transparency was adjusted. By using drop shadows I was able to get the feeling of the individual elements floating above the background, which is something that would have been very hard to do pre-Mac.'

'On the spinet, which has many small areas of colour, I masked out the black pixels I didn't want to change using the Magic Wand tool and the Select Similar command then painted on the colour before adding a bright yellow outline to emphasise the fluid shape.'

TECHNIQUE

Using images such as the Swing band and three tenors enabled Shorrock to get a feel of movement and sense of dynamism in the illustration.

Adding a white 'halo' to the viola by copying the layer, expanding the selection by six pixels, filling with white, then applying a 12-pixel Gaussian Blur filter lifts it off the page, making the image less flat and more three-dimensional.

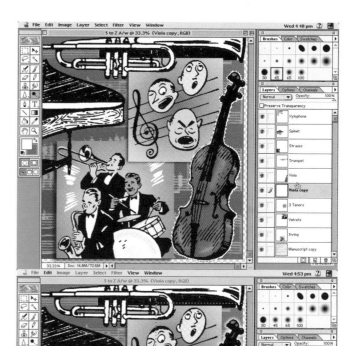

(technique) and KIT ᵥ

'My technique hasn't changed that much, it's still basically approaching a subject using a juxtaposition of found images to make an interesting visual "colouring book". With collage I have so many options, so many ways of illustrating an idea,' says Shorrock. One advantage to his technique of using black and white line images with flat colour behind it means they survive the fairly crude printing techniques of newspapers: 'You get inks varying, saturation, dots gain, paper expansion, shifted registration and so on with newspapers, and my technique survives those things because it uses strong basic elements,' explains Shorrock.

HARDWARE

> Apple Macintosh Performa 630, 36Mb RAM
> Epson GT-6500 Scanner

SOFTWARE

> PhotoShop 4.0

'THE CHALLENGES PRESENTED BY THE SERIES I THINK PRODUCED SOME OF MY BEST WORK, BECAUSE THE HARDER THE PROBLEM THE BETTER THE SOLUTION.'

PAUL SHORROCK
31 WYATT PARK ROAD, STREATHAM HILL, LONDON SW2, UK
TEL: +44 181 674 4420 FAX: +44 181 674 4427
E-MAIL: p.shorrock@mail.bogo.co.uk

CLIENTS INCLUDE: A&M Records; BBC; British Telecom; Channel 4; DDA; DMB&B; EH6; *Financial Times*; IBM; The British Council; *Esquire*; *Time Out*; *The Sunday Times*. [SEE PAGE 152]

Terry Colon is quite probably one of the world's best-known (and loved) on-line illustrators — a subjective view but one shared by much of the on-line community. For the past two years his work has graced the pages of the satirical column, Suck (at www.suck.com), one of the web's longest-running columns which is updated every weekday.

Suck has won numerous awards and a loyal following (no mean feat in the fickle and fast world of cyberspace). It's extremely well-known among the on-line community, for while it has spawned many imitators and homage sites, it is still acknowledged as having the best writing on the best subjects — with the best illustrations. And the majority of these are by Terry Colon, who to date has done well over 1,500 cartoons for the site.

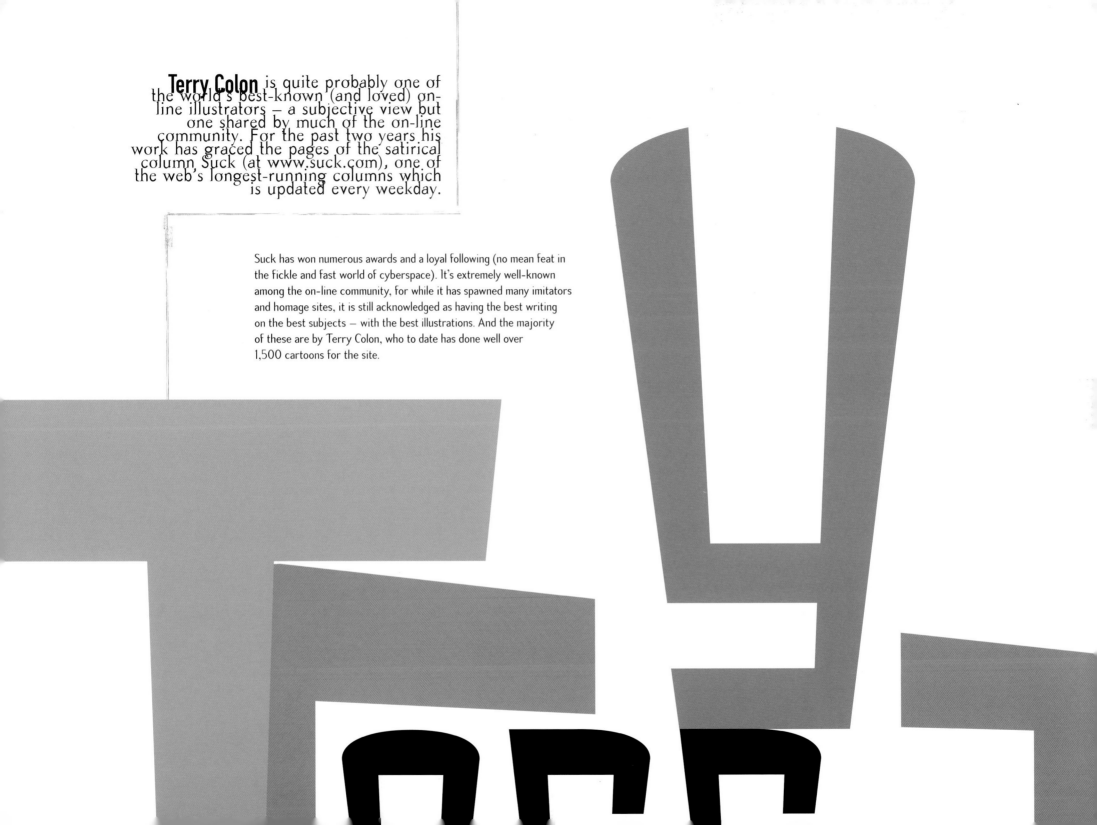

TECHNIQUE 'I take style and humour cues from cartoons and animation of the 1950s and early 1960s,' says Colon. Here he illustrates the question of expressing loyalty and fidelity to the brand in The Whoppers vs. Big Mac war.

Last year Joey Anuff, a Suck co-founder, proposed the idea of a Suck book to Wired Books (a part of Wired Ventures, publisher of *Wired* magazine). As Wired Ventures already owned the Suck essays, repackaging them for print made a lot of sense, allowing Suck to reach an audience beyond the digital world: 'Some [most] people still read type on paper and not the web,' says Colon. 'Since I had already illustrated a number of the articles, and I think they liked my work, I was asked to design and illustrate the book,' he adds.

(background)
BACKGROUND ^

Because the book is about the essays rather than the illustrations, it was decided to use the imagery in two-colour rather than the Day-Glo or eight-colours *Wired* tends to use in its more design-specific or graphics-orientated output. Here Colon talks about his work in general and the process of translating his on-line images to print.

NikeTown Crier
by St. Huck

Suck Prediction for 1997
http://www.suck.com/predictions/prediction10.html
"We dis and marketing giant Nike, after establishing a level of media saturation and mindshare so great as to make the use of either copy or a phenomenon identification in its ads redundant, within 1997 shed its last remaining vestige of symbolic language. Dumping the 'swoosh,' the shoe/sport/lifestyle purveyors declare the three corporate logo to be a 'space.' A multimillion-dollar Wieden and Kennedy campaign, containing an lavishly landmagazine inserts and television commercials consisting of sixty second segments of white noise, has unprecedented success; polls show seventeen to twenty-four-year-olds overwhelmingly identify pictures of white walls, unused tissue, and clear blue sky as Nike. In December, the company is named Advertising Age's 'Marketer of the Year' for the second year in a row. Shortly thereafter they are named God."

HIGH above Bryant Street in San Francisco, there's an eye-catching billboard. More flag than ad, it features no text and no photographs – just a black background, a red oval, and that ubiquitous white swooshette. Like the twentieth century's most notorious logo, from which it appropriates its color scheme and bold graphic style, the Nike billboard is a highly effective piece of iconography, galvanizing teenage thugs and suburban housewives alike with its symbolic magic. It announces the debut of Nike's retail presence in Union Square, but staring up at it, it's easy to imagine that the company's new chain of stores, extravagant as they are, are mere prelude to a far grander vision of corporate sovereignty.

Even wearing Air Jordans, it's a giant leap from NikeTown to NikeState, but Nike CEO Philip Knight, like any multibillionaire, must entertain at least occasional dreams of the status that statesmanship might afford him. After all, isn't fashion just fascism with more emphasis on uniforms than ammunition? Every time a new NikeTown opens, the faithful come in droves to pledge their fealty to Knight's vision: eight-year-olds break their piggybanks to buy overpriced wristbands; aging Yuppies dress golf cap to walking shoe in the emperor's new clothes. The patriotic equivalent of consumer devotion is increasingly rare today; even the militia movement can't match the brandinistas for numbers. Wandering the retail maze of the nation's NikeTowns, passively absorbing the totalitarian adspeak that adorns the walls, they wait for their call to arms, restless and dutiful.

"Mall of the Wild," by Hans Eisenbeis
http://www.suck.com/96.12eisenbeis/96.12eisenbeis.html
It may be hard to believe, but the Mall of America isn't just a metaphor; it's also an actual mall. But while Hans Eisenbeis, in "Mall of the Wild" on Feed, finds a profound sense of wrongness in the Minneapolis consumer cavern's Rainforest Cafe, auto dealership, Knott's Camp Snoopy, and FAO Schwartz Barbie Store, one still shouldn't lose sight of the phenomenon's far-flung analogs. Take the Plaza de Las Americas in San Juan, Puerto Rico. Also the size of a medium-sized hamlet, Plaza de Las Americas has recently found a novel method of synergizing its medical office annex with its sprawling retail space: beepers. Rather than flipping through the latest issue of ¡Hola! in the waiting areas, patients are given beepers and invited to shop until they're called. For all the talk of the anxiety-producing effects of modern consumer capitalism, shopping still tends to be less stressful than oral surgery.

Isn't it time, finally, to fully implement the corporation, the real Mall of America? As long ago as 1946, Peter Drucker was declaring the corporation America's representative social institution; in the fifty years since then, corporate influence upon our lives has become so routinely all-pervasive it's hard to remember it wasn't always that way. Until the late 1800s, though, corporations were chartered only for specific purposes and durations, with additional limits on land ownership and capitalization; now all it takes to start one is ten spare minutes and a few hundred bucks.

Today, the few people who dream of resurrecting those original restrictions are either dismissed as purveyors of paranoyzing cant or totally ignored. As for the rest of us, well, except for the most egregious displays of multinational malice, all is permitted. That's how it goes when you're in love – and really, which of the following institutions evoke the strongest feelings of ardor these days: Nike, Starbucks, and Nintendo, or Congress, the Executive Office, and the FBI? There's certainly no one building fan pages for Louis Freeh.

As much as we'd like to declare ourselves citizens of our favorite

brands, most corporations are probably somewhat less inclined to have us. Even if citizenship were invitation-only, corporation-states would still lose out on the hundreds of billions of dollars available to them now as corporate wealthfare. There'd be no more taxes, but also no more subsidies, bargain-basement land deals, or overseas co-op marketing campaigns. And without the divertissement of political scandal, public scrutiny of corporate behavior would likely intensify. There's a reason Clinton attracts more corporate support than any previous Democrat; his virtuoso facility for courting federal investigations makes him an excellent media baffle. With the press so engaged in deep-throating the details of botched blow jobs, there's less bandwidth to spend on corporate snow jobs.

On the other hand, given the current climate of consumer indifference to corporate iniquity, maybe baffles like Clinton aren't really necessary anymore. Outside NikeTown's San Francisco grand opening, activists protesting the company's employment practices in Indonesia were met with apathetic shrugs. What was it the ordinary Germans said in response to Hitler's diabolical directive: just do it? To assuage the few reporters who've shown more interest in sweatshops than shopping for sweats, Nike recently hired freelance Samaritan Andrew Young to put a positive spin on the situation – but in the land of NikeState, that kind of corporate rhinoplasty would be superfluous. Dissenters would be banished to the Birkenstock Nation.

Deliverance from government regulation and PC activism would certainly be enough to make some corporations embrace the new world order. Gigantic companies could forsake their strained attempts at stealth marketing and return to the good old days of honest, aggressive addiction cultivation. McDonald's could threaten rainforests with impunity, and cosmetics companies could start tarting up bunnies like drag queens again.

Of course, too much corporate self-interest would still have dire consequences; consider the case of the Republic of Cuervo Gold, a small

Ending Corporate Governance
http://www.artcrimes.com/corporation/
"It is essential to understand how corporations prior to the Civil War were legislatively defined, so we may better appreciate what we can discern and make use of today using the sections still present in our state constitutions – as well as restating and strengthening in favor of nature, citizens, and communities many sections that have been repealed by corporate agents seeking to make incorporation laws more 'corporate-friendly' – to one allow corporate authority, and mis-state the authority of we the sovereign people."
Sure, it bears an eerie resemblance to the creed of the People's Republic of Texas, but secessionist doctrine often follows well-worn tenets: appeals to the letter of the law, often of previous centuries; tightly delineated populist jingoism; and, always, an alternative rewrite of history, often not wholly unfounded. The authors of this digital broadside incite toward a revolt against the corporate plutocracy. While the plausibility of their liberated citizen's revolution may be on par with a flat-earther over throw of NASA, they do provide a fascinating historical retrospective of the history of the corporation, reminding neophyte sovereigns that prior to the mid-1990s, corporations had limited durations, limited allowances on ownership of land, limited capitalization, limited purpose, and unlimited liability.

SALES HE

big, dry, and Nevada-, despite its huge surplus of smarmy guy-like Dude,
the Republic is languishing – apparently no one can exist on (or stand)
tequila, sand, and Cuervo Gold Ambassador Dan Cortese for more than
a few hours. To succeed in a world of corporation-states, strategic
alliances would be more necessary than ever.

How such partnerships all sort out, however, is ultimately inciden-
tal. The important thing about the evolution from nation-state to corpor-
nation is how this change would reinvent our lives with meaning. With
religion reduced to little more than vaporware and PR, and patriotism a
mere marketing technique for celebrities who can't sing or act, we have
few real opportunities to seriously express our belief in anything. We
love our brands, yes, but what can you do to show that love except buy
lots of crap and maybe make a Snapple commercial? A world of corporation-states would inevitably present more meaningful ways to prove
faith: Whopper vs. Big Mac? Now that's a war worth fighting.

72

'WHO CARES HOW A THING IS DONE? LIKE MY DAD (WHO GOT ME STARTED IN THE BUSINESS WORKING IN HIS STUDIO AS A KEYLINER/PASTE-UP GUY) USED TO SAY: "IF IT LOOKS RIGHT, IT IS RIGHT".

THE PROJECT >

For the publisher, the idea of the book was quite straightforward:
take a number of Suck articles with their illustrations and reproduce
them in print. But while most of the illustrations which were to
appear in the book had already been created by Colon, translating
them for the book required reworking the original pictures: 'I had to
clean up the line work and re-scan the original drawings (done using
marker pens on layout paper) at a finer resolution. I then converted
them to vector art using Adobe Streamline and assembled, coloured
and edited the final images in Adobe Illustrator,' he explains.

SAIGON
SHOE
COMPANY

"Work Makes
You Free"

'When I draw for the web my images are fairly rough because, at 72 dpi, screen
display doesn't show crisp edges anyway. But in both cases I draw the art on
paper, scan it, convert it to vectors in Streamline, colour and edit it in Illustrator.
Then for the web there is the additional step of rasterizing the image in
Photoshop to convert it to a photogif. In the end they are completely different
pieces of art, the book version being a vector-based Illustrator EPS and the web
one being a pixel image photogif,' he continues.

TECHNIQUE

For a scathing and succinct
analysis of corporate USA, with its
reliance on sweatshops and
covert acceptance on the part of
an adoring public, Colon wittily
drew up illustrations which dealt
with the themes of tyranny, strat-
egy and exploitation explored in
the article.

MEDIA

CLINTON

Colon designed the layout of the Suck book as well as doing 150 illustrations for it. As it's not possible to have links on the printed page he designed the footnotes and side bars to be the print equivalent of links, and by connecting sections of text to side bars with lines and graphic treatments he aped highlighted text. 'After all, hot links are just electronic equivalents of side bars, footnotes, or references anyway,' he reasons.

TECHNIQUE

DEVELOPMENT
(development)

On all his work Colon starts with the manuscript, which he reads through once to get the gist of the piece: 'Since Suck covers a wide range of topics from politics to culture to marketing to everything digital, it means I am often drawing something new and different. Our contributing writers are first-rate, witty and not at all dry or technical, so it's entertaining reading them,' he enthuses. He then rereads, looking for specific ideas in the text that might lend themselves to illustration.

'I don't like to just take a bit of text and illustrate it verbatim, but rather I try to find an idea or image in the essay that I can give a new twist to. I often like to start with what may be a familiar image or a cliché and either change an element or add something unexpected. I don't focus on the process or where the ideas come from. The way I work is a lot more intuitive. I read the manuscript and something pops into my head as if by magic,' he explains.

'When I have a block I try to pick out something visual that's mentioned or implied in the text and put that element into a classic gag cartoon situation — the angry guy with a sign, stranded on a desert island, something falling on someone's head, wind-up key in the back, that kind of thing. When all else fails I just select a sentence and draw it as funny as I can make it.'

In designing the Suck book, Colon decided to loosely reflect the website: a narrow column of copy punctuated with illustrations. 'Primarily the book is about humorous essays so I wanted the design to look fun but not hinder readability,' he explains. He used ITC Century Schoolbook for the main text: 'My favourite typeface for body copy. It has a large x-height, looks good without being quirky or distracting, colours up nicely and is easy to read. The side bars and intros are in the Zurich family, and the display type Ad Lib, which compliments my drawing style very well,' he explains.

POOPY'S SCRATCH & SNIFF GROSS BOOK

POOPY

'WHEN IT COMES TO COMMERCIAL ART I ENTIRELY AGREE WITH THE GREAT CARTOONIST WALLY WOOD WHO SAID IT BEST: "NEVER DRAW WHAT YOU CAN COPY. NEVER COPY WHAT YOU CAN TRACE. AND NEVER TRACE WHAT YOU CAN CUT OUT AND PASTE DOWN." ONE MIGHT UPDATE THIS ADAGE TO BEFIT THE DIGITAL AGE, BUT THE MAIN SENTIMENT REMAINS AS VALID NOW AS EVER.

Crooked lines are funnier than straight lines.
Dotted lines are funnier than solid lines.
Light colours are funnier than dark colours.
Flat is funnier than perspective.
Ugly is funnier than cute.
Top-heavy objects are funnier thanbottom-heavy ones.
Really big is funnier than really small.
Small kitchen appliances are funnier than furniture. Always include either a
blender or, better yet, a toaster in every kitchen setting.

'Because screen displays are not very high-resolution, I am able to get away
with rough edges and so my drawings don't have to be that precise. I use a
mouse not a tablet. I've tried but I just can't draw on the computer with a stylus.
I use traditional methods, pen on paper, and then scan. It may not be the best
way for everyone but it works for me. Our maximum image width is 200 pixels
so I don't draw very large or complicated images. Details are lost on a 72dpi
monitor at any rate,' says Colon. Images are scanned at 400dpi for the website and
800dpi for the book. For both the website and the book, the images are converted
to vectors in Streamline, before being coloured and edited in Illustrator. Web images
are then rasterized in Photoshop to convert them to photogifs.

TECHNIQUE

To recreate the original website illustrations for use in the book, Colon had to to go back to the original drawings and clean up the line work before re-scanning them at a finer resolution. These were then converted to vector art using Adobe Streamline and assembled, coloured and edited in Adobe Illustrator.

INFORMATION

Last year, Terry Colon wrote and illustrated Suck's column for 7 November 1997 on the principals of comic drawing. It's well worth looking at the whole thing at http://www.suck.com/daily/97/11/08

'I'M INSPIRED BY EVERYTHING GOOD I'VE EVER SEEN AND LIKED BY ANY ARTIST ILLUSTRATOR OR CARTOONIST — FROM MARCEL DUCHAMP TO GARY BASEMAN TO TEX AVERY TO GUY BILLOUT TO ROZ CHAST TO ANDY WARHOL TO BRUCE McCALL AND MANY OTHERS TOO NUMEROUS TO MENTION.'

TECHNIQUE

'I try to colour my illustrations with a sense of humour influenced by the kind of gag cartoons you get in newspapers and magazines like *Playboy*, *The New Yorker*, *Mad*, etc. I attempt to make each picture a little gag cartoon that could stand on its own even if you didn't read the story. And hopefully all the more amusing if you did.' explains Colon.

HARDWARE

> OFFICE: POWER MAC

> APPLE MULTISCAN 20" MONITOR

> HP SCANJET

> HOME: POWERCENTER (MAC CLONE)

> 20" SONY TRINITRON MONITOR

> UMAX SCANNER

SOFTWARE

> ADOBE STREAMLINE, ILLUSTRATOR AND PHOTOSHOP. QUARKXPRESS 4.0.

> DESIGN AND SHARPIE MARKER PENS, LAYOUT PAPER

> AND, SOMEWHAT SURPRISINGLY FOR AN ON-LINE ILLUSTRATOR, AN 'IOMEGA ZIP WHICH I USE TO TRANSFER WORK FROM HOME TO OFFICE AND VICE-VERSA AS I AM NOT ON-LINE AT HOME.'

TERRY COLON
554 LOMBARD STREET, SAN FRANCISCO, CA 94133, USA
TEL: +1 415 276 8556/415 835 2422
E-MAIL: terry@www.suck.com

CLIENTS INCLUDE: *Fortean Times*; Visable Ink Press; *Cracked* Magazine; Applause!; Unix World; Home Office Computing; Ford Motor Company; *Infoworld* Magazine; Wired Ventures. [SEE PAGE 156]

ESSAY BRAZIL

While it's arguable that the US leads the field in digital art and illustration, work being done throughout Europe by freelance illustrators is recognised as bringing a sensibility and style influenced by and referring to classical art which is often lacking in New World work, which tends to draw more on contemporary influences. 34-year-old Italian digital artist and illustrator **ALESSANDRO BAVARI** is a prime example of a contemporary illustrator who uses the computer to 'create images which are formed through an unique fusion between traditional techniques and computer graphics'.

Bavari's training in scene painting and history of art at the Accademia delle Bella Arte di Roma gave him a strong grounding in classic techniques such as oil painting, various engraving styles, watercolours and photography, and with this classical background, he 'felt the need to experiment with a new visual language which would sum up all my artistic experiences'. Since leaving college he has worked as a freelance artist and illustrator, mixing his college-taught skills with self-taught computer skills to create work for clients such as Adobe, *MacUser*, McKann Erikson and numerous Italian museums, galleries and publishing houses. He has also exhibited in competitions and exhibitions worldwide.

'I THINK THAT MY WORK NOT ONLY HAS A EUROPEAN SENSIBILITY BUT MORE SPECIFICALLY AN ITALIAN ONE, INFORMED AS IT IS BY THE WORK OF GIOTTO, BEATO ANGELICO AND MASACCIO, BUT ALSO BY THE DELIRIUM OF BOSCH, BY GOTHIC ART AND BY THE FLEMISH PAINTERS.'

(background)
BACKGROUND >

Without distancing himself from the classical canons, 'indeed, quoting in my work the symbolic importance of the sacred images represented in the 14th- and 15th-century work of artists such as Giotto, Paolo Uccello and Piero della Francesca,' Bavari recently began to elaborate on his personal techniques, 'using industrial and organic products from nature before incorporating photographic manipulation then digital manipulation'. Last year he started work on a project which would draw its inspiration from classical mythology, resulting in a series from which the work shown here, Adolescent Mercury, is taken. 'I felt a desire to work on something personal, a series of art which would represent the current state of my artistic research and experiences,' says Bavari.

(the project)

Looking through a book on the work of Flemish painter Jan van Eyck, Bavari was struck by the 1439 'Portrait of Margaretha van Eyck', 'particularly by its solemnity and aesthetic composition', says Bavari. This work was to inspire the composition of Adolescent Mercury strongly.

Bavari began to consider the individual elements for the piece: 'My favourite materials are those sea-swept objects tossed on the beach after a storm; worn and shaped by the elements, I feel they are surrounded by a supernatural aura. While I rarely remove them, I do take lots of pictures of them,' he says. 'I also love taking photos in natural science and history museums where each object, being outside its natural context, immediately takes on an appearance of metaphysical, almost magical beauty,' he adds.

INSPIRATION

This Jan van Eyck painting from 1439, 'Portrait of Margaretha van Eyck', was the original inspiration for Adolescent Mercury, which forms part of a series of digital work Bavari created based on classical mythology: 'I love the solemnity and composition of this Van Eyck piece, and it obviously stayed with me because I did the Mercury piece quite a while after I saw it, but it inspired me nevertheless,' he says.

alessandro bavari | electronic workshop

PHOTO ARCHIVAL — Bavari shoots objects all the time and archives them for future use. This tree was photographed in Spain some years prior to it being incorporated into Mercury, where not only is it used in its whole form but the pattern of the foliage is superimposed over Mercury's head.

DEVELOPMENT
(development)

'My idea from the outset was to represent Mercury in a non-traditional, unusual way,' says Bavari. So the main torso of the piece is formed by a doll he found on a beach, representing the frivolity and impermanence of youth, and to this Bavari began to add other elements such as the photos of the tree, the swan and the wings, and some old watch gears and cogs. 'Some of these were scanned in directly, others were photographed, and I used a painted background on which to compose the elements,' he explains. Although the doll's head had eyes, Bavari wanted to add real life to the piece and so substituted them with a friend's eyes photographed specifically for that purpose.

DIRECT SCANS — These cogs and gears taken from an old watch were directly scanned in by Bavari, who likes to juxtapose imagery from nature and technology or science, mixing organic, living objects with industrial ephemera.

TRADITIONAL TECHNIQUES — Bavari is keen to use his classical art training in his images, so will often create backdrop scenes for his digital work from oil paints, engraving, water colours and other traditional methods. For this background, 'using hands, brushes, nails, combs and rags', he painted oils directly on to a sheet of zinc which was then pressed on to watercolour paper that had been soaked for an hour.

'Generally I always try to connect my subjects and elements in a special ambient which is ambiguous and undefined, so I like to create backgrounds painted in oils and add layers and mix them with scans of negatives which have been chemically treated, and beyond that ripped or mutilated in some way,' explains Bavari. 'Whereas in the past I worked and experimented a lot with traditional photography, arriving at interesting techniques in things like camera obscura, now I tend to concentrate on the level of experience I've gained in the past few years, and I do all that experimentation with digital techniques on the computer, an area where the absence of materials reigns but where everything you do is exactly what you would have done with the old methods,' he says. Elaborating on that, he says: 'I mean the mental, artistic process and the organisational criteria required in planning a piece of work. The big difference of course is that you lose an intermediate stage, the manual one, but that's not important.'

'WORKING IN PHOTOSHOP LAYERS AND EMPLOYING THE TOOLS EFFECTIVELY, THE DETAIL OF THE WEAVING AND TEXTURES AND HOW YOU JUXTAPOSE THE LAYERS CAN RESULT IN AN AUTHENTIC AND REALISTIC PIECE WHICH COULD HAVE BEEN GENERATED NOT WITH A COMPUTER BUT USING MANUAL TECHNIQUES.'

TORSO

Bavari found this body of a doll on the beach following a storm and knew it was perfect for his torso of Mercury. He substituted the eyes with photos of real eyes, wanting to add an element of real life to the composition.

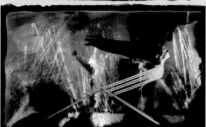

(technique) and KIT ∨

'Having assembled the piece in my head, I turn to the computer to start realising the idea in Photoshop,' says Bavari. Each element is precisely scanned and made into a single layer, and sometimes the same object is duplicated into more layers which are differentiated through various tool options relative to those used on corresponding layers.

'Often I am working with 50 layers, and through moving these layers and refining them I begin to form the composition of the piece,' says Bavari. Generally he converts his transparencies and photos into greyscale, using personalised settings to 'arrive at effects that were previously only achievable through the use of chemicals'. Through extensive experimentation, Bavari has created a personal library of colour effects, 'synthesising things like the tonal effects in metals and trying to capture the nature of an object or colour through the textures and tones which characterise it,' he says.

ELEMENTS { While the swan was an embalmed exhibit Bavari photographed at the National Museum of Prague, the wings, used on the sides of Mercury's head, are those of a dead bird he found in a street gutter. The piece of plastic was a formless object which he found on a beach and decided to use as a decorative element in the piece.

'The image is then converted to RGB in order to utilise all the parameters of control on the tones and colour, such as levels, curve, colour balance, selective colour etc. Along with commands such as calculations and apply image, which perform mathematical calculations on the pixels, you can achieve effects which are unexpected, but pleasingly so,' adds Bavari.

'Once the composition is complete, I start to mess it up or add dirt to the image by superimposing textures and adding dust and scratches, along with blurring and contrasting details. I try not to go overboard on this, letting it develop in an intuitive or fluid way, I think in a painterly way.'

'WHAT I LIKE ABOUT THE COMPUTER IS THE WAY YOU GO DIRECTLY FROM THE IDEA TO THE FINISHED THING, A *FAIT ACCOMPLI.* LOSING THAT MANUAL INTERMEDIATE STAGE DOESN'T DIMINISH THE WORK, AS THE ARTISTIC, ORGANISATIONAL PROCESSES ARE THE SAME WHATEVER TECHNIQUES YOU'RE USING.'

Finally, Bavari outputs his work on to photographic paper up to 80 x 120cm, using the Lambda system, 'a kind of gigantic digital enlarger which writes the image directly to the paper using lasers based on the RGB colours. This can then be developed or fixed according to the traditional methods, in the classic chemical baths, thus returning to the preciousness of a material which has been in use for more than a century, like photographic paper,' concludes Bavari, obviously happy to once more pay homage to the methods which he feels the computer can successfully emulate but never replace.

HARDWARE

> MACINTOSH PPC 9500/120 WITH 2.3GB HARD DRIVE AND 160MB OF RAM
> UMAX SCANNER
> KODAK DC120 DIGITAL CAMERA
> IOMEGA JAZ DRIVE

SOFTWARE

> PHOTOSHOP 4.0
> PHOTOCD PRO AND STANDARD

ALESSANDRO BAVARI
VIA CARDUCCI 7, 04100 LATINA, ITALY
TEL: +39 773 696 828
E-MAIL: abArt@bmnet.it

CLIENTS INCLUDE: McKann Erikson; Romberg Contemporary Arts; Fiumara d'Arte; WWF. [SEE PAGE 142]

From a small New York studio housing one intern and designer, Hjalpi Karlsson, Austrian-born **STEFAN SAGMEISTER** has the dream job of designing CD covers for the likes of Lou Reed, David Byrne and The Rolling Stones. Virtually all of his clients are record companies or bands, 'partly because we love music and partly because we get a much bigger kick out of meeting and working with musicians we hugely admire than working with marketing managers,' says Sagmeister.

After working with esteemed design group M&Co in New York and as creative director at the Hong Kong office of ad agency Leo Burnett, Sagmeister set up Sagmeister Inc. in 1993. Unusually, he has firm plans to ensure the studio stays small: 'It means less meetings, less administrative work and less overheads, all of which allows us to pick and choose projects as we want,' says Sagmeister. 'It also means we all do a bit of everything, which makes our work more interesting,' he adds.

(background)
BACKGROUND<

With a portfolio that includes just about every major record label, it was no surprise that when The Rolling Stones had to choose a designer for last year's album, 'Bridges to Babylon', they decided on Sagmeister Inc. Unusually, it was the band rather than their record company, Virgin, who commissioned (and paid) Sagmeister, making things a little easier for the designer: 'It meant fewer people were involved in the design decisions, but Virgin were very good in their input, for example they teamed us with Debbie Kara, who was responsible for the production of the cover. She was fantastic and sorted out a number of problems for us,' says Sagmeister.

When Sagmeister first met with Mick Jagger to discuss the album, there was no title or music available. Jagger was keen that the cover had some relationship with the stage set Mark Fisher had designed for the world tour, which incorporated different cultural and historical references and elements in a fluid way. Sagmeister agreed in principal but decided to make the connection more spiritual than literal.

'WHEN CDs FIRST CAME OUT I WAS BITCHING WITH THE REST OF THEM ABOUT HOW IT WAS TERRIBLE FOR ALBUM COVER DESIGN, NOW I THINK THE 12" COVER IS REGARDED AS THE BASTARD CHILD OF THE CD COVER. THE CD COVER WE DID FOR LOU REED'S 'SET THE TWILIGHT REELING' ALBUM COMES IN THIS DEEP BLUE CASE WHICH IS KEY TO THE DESIGN, BUT THE RECORD COMPANY DIDN'T WANT TO SPEND MONEY ON THE VINYL VERSION, SO THEY JUST SHOT A BLUE ALBUM COVER, WHICH MISSES THE POINT COMPLETELY.'

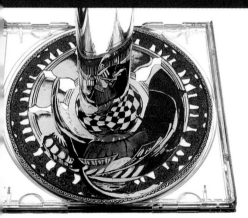

DUMMIES } After an initial meeting with Mick Jagger – but before a title or the music were available – Sagmeister presented dummies and sketches of four different directions. Elements of these – such as the silver reveal – made it to the finished design.

(the project)

THE PROJECT
v

When Jagger and Charlie Watts met up with Sagmeister in London a month after the initial meeting, the album had been named, 'Bridges to Babylon'. At Jagger's suggestion, Sagmeister started looking at Babylonian art in books and also paid a visit to the British Museum's collection. 'There was one piece there which really stood out, this statue of an Assyrian lion, so I took pictures of it and Mick and Charlie really liked it,' says Sagmeister. 'For the next week I sketched and what began to emerge was a move away from this static, three-dimensional representation, which tends not to work very well as a two-dimensional representation, to a more animated image reminiscent of those 1970s' science fiction book jackets,' he continues.

Sagmeister had also decided that the lion in a heraldic pose would reflect the historical narrative of the stage set, while a silver reveal on the cover would echo the silver curtain of the set. This was inspired by a trip he had made to India a few months earlier: 'They have these garish, four-colour images overlaid with gold-stamping, and I liked the idea so much I decided to incorporate it into a project — luckily it suited this one perfectly,' recalls Sagmeister.

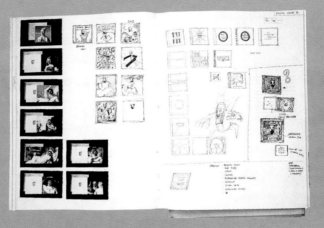

DEVELOPMENT } Sagmeister spent a week presenting daily ideas to Jagger based around the album's title, 'Bridges to Babylon'.

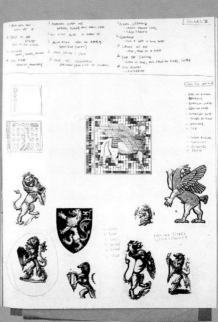

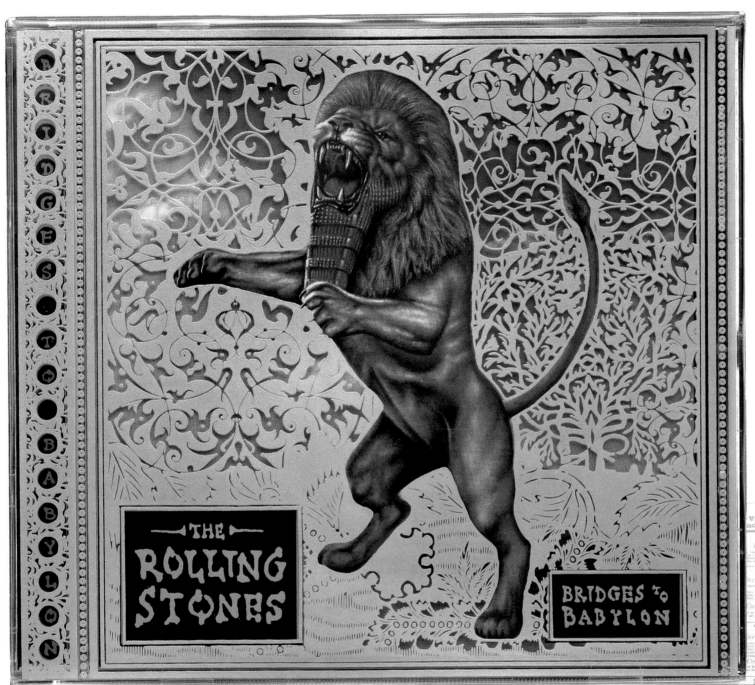

THE CASE

The final cover features the work of three illustrators, a photographer and Sagmeister Inc. The pull-off silver reveal case echoes the stage set designed by Mark Fisher, which features silver curtains and different cultural elements which are subtly incorporated in the patterns on the case: 'the lower part of the pattern is Japanese, the top left medieval German and the middle portion on the right is Moroccan,' explains Sagmeister. Type is scanned from hand-drawn sketches by Sagmeister.

VARIATIONS { After deciding on an Assyrian lion in a heraldic pose, influenced by a statue at the British Museum, Sagmeister presented numerous different colour and background variations.

Back in New York, Sagmeister began to develop the various elements. He commissioned Kevin Murphy, with whom he has worked on numerous projects, to illustrate the lion. 'We gave Kevin a pretty tight brief because we had such a strong idea of what was needed,' explains Sagmeister. 'He came up with an oil painting which had to be a lot more intricate than a normal illustration would because, once scanned, the same file was going to be used for everything from the CD cover to huge billboard ads and posters,' he adds.

'All that was required of me was creating the actual lion and then painting the concept with my personal flavour,' says Murphy. 'It went through several changes before being approved. In the original concept the lion had a more human face and a braided cap, we also played around with the level of muscularity,' he adds. One problem was working the lion out to not only look good as an illustration, but to still be recognisable in silhouette. Sagmeister's original sketch fit both requirements, but when Murphy began working with pictures of a real lion, things didn't fall into place so simply. 'As each of the drawings was completed, small corrections were made (moving it more in the direction of a real lion) until Stefan and I were both comfortable with the look. The finished version of the lion is very close to Stefan's original design,' explains Murphy.

DEVELOPMENT
(development)

The backdrop was created by Gerard Howland at The Floating Company, a company introduced to Sagmeister by Jagger. 'They had done costumes, props and the backdrop for the stage set, so it made sense for them to do the backdrop for the CD cover. We'd already rejected a few desert scapes — like Death Valley-type scenes littered with rubbish — but Gerard gave us pretty much exactly what we wanted,' says Sagmeister. Final illustrations, vignettes of Babylonian columns, were drawn on computer directly on to Howland's backdrop by Alan Ayers.

Typography came from tiny pencil sketches drawn by Sagmeister. These were scanned in and the contrast changed to ensure no half-tones. 'After my initial drawing, which was smaller than appears on the cover, I tried out a number of variations to tidy it up, but seemed to move away from what I was trying to achieve, so ended up using that first one because it best captured a sense of looser "sloppiness",' he explains.

By now the deadline was looming but no decision had been made on track order, which forced a happy accident. Originally Sagmeister had intended the silver reveal to be printed on to the plastic CD case, but the track order problem meant that this couldn't be done and Sagmeister decided to make the silver 'curtain' a separate slip case: 'At the time I really wanted them to be part of the same thing, but now I actually think they work better as separate elements,' he comments. 'Debbie Kara was instrumental in getting this right,' he continues. 'She decided to get a prototype shrink-wrapped, which luckily showed up the fact that the slip case was $1/64$th of an inch too big!'

APPLICATION { As the same artwork for the cover was to be used on eight-storey-high billboards, the lion illustrated by Kevin Murphy had to be incredibly intricate and detailed.

COMPS { 'We comped up the inside but then decided against it – too mystical – and redesigned it with the simpler desert scape shown on the sketches and final cover.' says Sagmeister.

'WHAT I LOVE ABOUT DOING CD PACKAGING IS IT HAS A GREATER VALUE TO THE USER THAN ANY OTHER FORM OF PACKAGING. LOTS OF PEOPLE EXAMINE IT IN-DEPTH TO SEE WHAT IT REVEALS ABOUT THE BAND AND THE MUSIC, AND UNLIKE OTHER PACKAGING IT DOESN'T GET THROWN AWAY SO HAS THIS WONDERFULLY LONG LIFESPAN.'

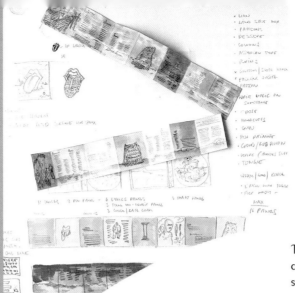

The size of the files for the three illustrators and photography by Max Vadukul caused a number of headaches for Sagmeister's studio, but a huge hard drive and some 'pretty impressive RAM' allowed them to manipulate and realise the project.

Sagmeister is very happy with how the finished product looks in stores, but admits: 'Maybe I would like the concept to have a bit more dimension. On some of our other projects the case is one thing and the inner sleeve offers something more, whereas with this it's pretty much all in one direction. But it still works and still has a real value,' he ends thoughtfully.

(technique) and KIT >

HARDWARE	SOFTWARE	ILLUSTRATION
> Apple Macintosh 9500/500 – 'with lots of RAM!'	> Photoshop 4.0	> Windsor & Newton Oil Paints
> ScanMaker 2SP	> Illustrator 7.0	> Windsor & Newton Series 7, #1 Brushes
	> QuarkXPress 4.0	> Cold pressed illustration board

SAGMEISTER INC.
222 WEST 14TH STREET, NEW YORK, NY 10011, USA
TEL: +1 212 647 1789 FAX: +1 212 647 1788
E-MAIL: Ssagmeister@aol.com

CLIENTS INCLUDE: Graphics and packaging for David Byrne; Energy Records; Aerosmith; Warner Bros. Records; Lou Reed; HBO Studio Productions; Viacom. [SEE PAGE 154]

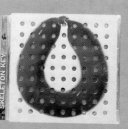

Until 1994 type designer, illustrator, designer and comic book artist **RIAN HUGHES** produced a hugely prolific range of work on a humble Macintosh IIcx. This range includes more than a hundred typefaces, including the well-loved Knobcheese; diverse projects for publisher 2000AD which include contributions to *Judge Dredd* and a revamp of 1950s' icon Dan Dare; a *Love and Rockets* novel cover and a reworked Batman; illustrations for MTV, countless editorial projects, packaging and album cover designs, the logo for *Forbidden Planet*, the shopfront for Fiell and more recently an animated safety video for Virgin Airlines.

While studying graphic design at the London College of Printing, Hughes discovered European comics, in particular the work of artists such as Serge Clerc, and turned to this field after stints at lifestyle and fashion magazine *i-D* and a short-lived stint with Tomato's John Warwicker at Da Gama. From this he expanded into virtually all areas of graphic design, illustration and type design. All his computer skills are self-taught: 'It allows you to pin down the bare bones, explore composition (shape) and colour a lot more easily, and they're the two things that interest me the most,' he says of the computer.

'My comic work has evolved from the very representational, precise work I did on Dan Dare in the 1980s, to the more stylised representations I use on *The Guardian On-Line* strip. Here I reduced the form to a very cartoony, flat image, with forced or no perspective, because I was becoming much more interested in the shapes made by the compositional elements. This held my attention more, and these abstract ideas I wanted to explore lent themselves more to illustration than comics. There's also a much higher turnover of ideas because as one-offs they're so much quicker to do than a lengthy serial,' explains Hughes.

(the project)
THE PROJECT
∨

Hughes had a very specific idea of what he wanted before the presentation: 'What I had in mind was "Top of The Pops" in the 1970s, when their countdown would use illustrative representations of the numbers – two skyscrapers forming the number 11 comes to mind. As "Box Tops" is on every day I felt I needed to create a sequence that would be full of busy details so as to withstand repeat viewing, I wanted something that was pretty mad and full-on. I knew I wanted a structure that logically produced the countdown in one continuous shot, so I came up with the idea of a zoom which starts from outside a dome city on the moon and goes right through people's living rooms, over swimming pools and so on, ending with a big explosion of stars, spirals and flowers. I was after very bright, brash pop images. On seeing the storyboard they loved the idea, so then it was just a matter of making it all work,' he says.

'WITH MY WORK I ALWAYS TRY TO STRADDLE THE LINES BETWEEN ILLUSTRATION, TYPOGRAPHY AND DESIGN.'

(background)
BACKGROUND
∨

Early in 1996 cable station The Box invited Hughes to design the opening credits to 'Box Tops'. The Box is a music channel similar to MTV but controlled by the viewer, who phones in to request particular videos, and 'Box Tops' is the weekly show comprising clips of the ten most requested videos on the station. The channel wanted the programme's title to be redesigned, but the only prerequisite for the opening sequence was a graphic countdown from ten to one (a structure similar to 'The Chart Show's'), lasting no more than 20 seconds.

AESTHETICS

'With "Box Tops", on the one hand you have a commercial purpose, to get people interested in the TV show, but on the flip side of this you're dealing with very abstract ideas – shapes and colours that go together in certain ways better than other ways, and why it is that they do so. This is the visual language inherent in nature you're trying to explore, and I suspect it's there whether we have a culture to describe it or not. Sometimes it's almost a relief on a project to have that front end of what it's for determined by the client – the aims and objectives are already defined.'

STORYBOARD

The storyboard created by Hughes is remarkably similar to the finished film and uses the narrative of a long zoom for the client's requested ten to one graphic countdown.

6. Cut back to saucer. Saucer exits frame at right. Behind saucer is a building inside the dome with a window in the shape of an 8.

DEVELOPMENT (development)

From presenting that one concept, the finished sequence 'came out 99 per cent what I had imagined', says Hughes. The process was a straightforward one once Hughes had done a very detailed storyboard, which he created using pen and paper. 'Generally I don't work on paper anymore, but with this it was quicker to actually sit down and storyboard it in traditional pen and ink rather than at this early stage draw it up on the Mac,' he explains. After this stage had been approved he produced the large number of drawings required for the final piece in Adobe Illustrator, including several different variations on the logo.

'I WORK VERY QUICKLY BECAUSE CLEAR THINKING IS THE MOST IMPORTANT THING, AND IF YOU KNOW WHAT YOU'RE DOING, THOUGHT IT THROUGH WELL, AND THE JOB HAS THAT INNER LOGIC AND CONCEPTUAL SYMMETRY THAT GOOD DESIGN HAS, WHERE EVERYTHING DICTATES OR RELATES TO EVERYTHING ELSE, THEN YOU'RE AWAY, AND SWEATING OVER IT FOR DAYS CAN SPOIL IT.'

LOGO ALTERNATIVES

'A lot of my fonts – about half of them – come about because I have a very strong idea of how a font could work for a particular project. How a font develops is based largely on the style of the font. For example, on a very rounded, curvy font I'll start with something like the letter "S" or "E", while with a more blocky font I begin with the "E", and the "S" comes very far down the line. But there are no set rules, because what letter best encapsulates the feeling of a font changes with each font.'

FINAL LOGO
The final logo chosen by the client and Hughes.

(technique) and KIT ˅

Having produced all the scenes in grouped layers, allowing Hughes to pull out elements and manipulate them as required, the next step was to animate them. Hughes worked with Phil Mann, then of DesignLab, now at prestigious SFX company Smoke and Mirrors, in digital effects package Adobe AfterEffects. 'Because After-Effects is another Adobe program, almost a bolt-on to Illustrator, I was able to take the drawings I produced and directly assign skeletons or transparencies to them and describe paths along which the various images would move.'

'This meant the finished animation contained my actual drawings rather than someone else's reworked interpretations,' explains Hughes. 'In that sense it's like working with an editor, deciding how to have objects perform, where and when to zoom, when and how to lay on the music track,' he adds. The music, a 20-second loungecore extract by The Ray McVay Sound called 'Kinda' Kinky', suits the sequence perfectly.

'I've only actually seen the titles go out once live, as I don't have satellite or cable. It's not intended to be some kind of high concept piece, it's meant to be loud, bright and fluffy. It's POP: in short, it's exactly what was wanted,' concludes Hughes.

'I THINK THE COMPUTER HAS EMPOWERED PEOPLE FROM ALL DISCIPLINES TO TAKE CONTROL OF A WHOLE PROJECT AND ALL ITS DISPARATE ELEMENTS, SOMETHING THAT WOULD HAVE BEEN VIRTUALLY IMPOSSIBLE TO DO IN THE PAST.'

HARDWARE

> APPLE MAC 7100/80

SOFTWARE

> ADOBE ILLUSTRATOR
> ADOBE AFTER EFFECTS

RIAN HUGHES
6 SALEM ROAD, LONDON W2, UK
TEL: +44 (0)171 221 9580 FAX: +44 (0)171 221 9589
E-MAIL: RianHughes@aol.com
WEBSITE: http://www.devicefonts.co.uk

CLIENTS INCLUDE: DC Comics; Forbidden Planet; MTV;
Mambo; Virgin Airlines. [SEE PAGE 153]

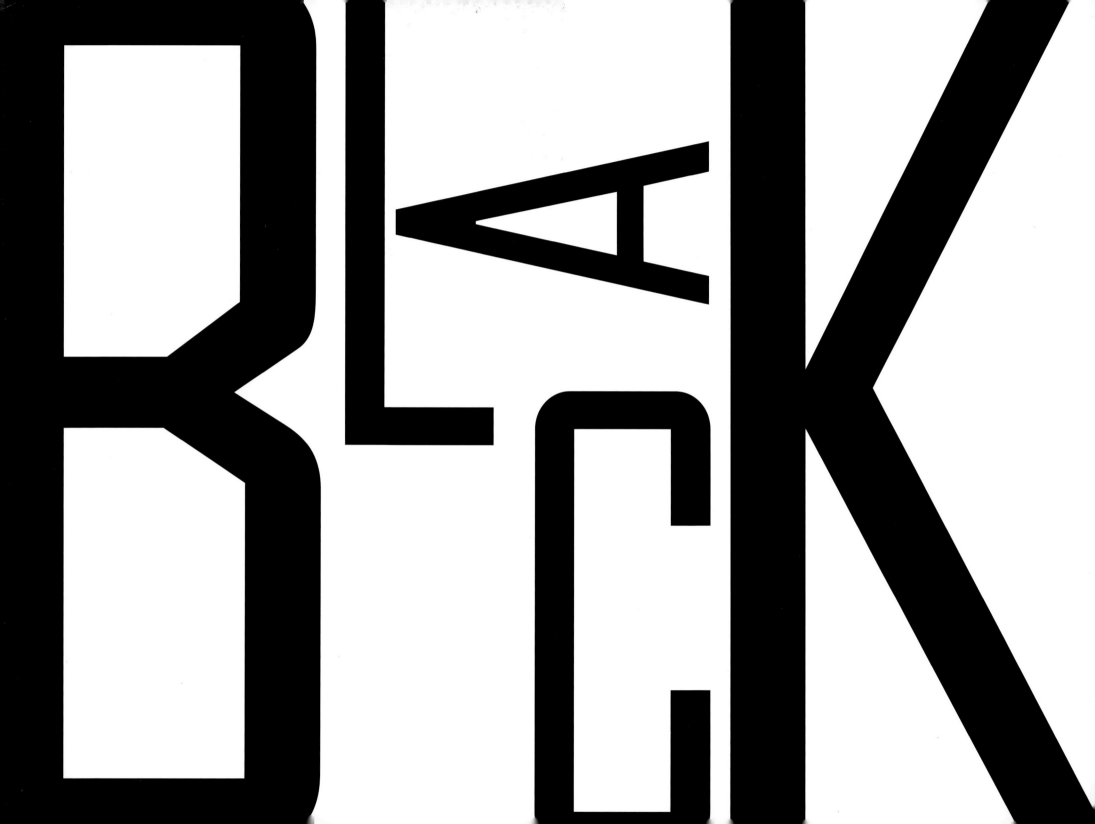

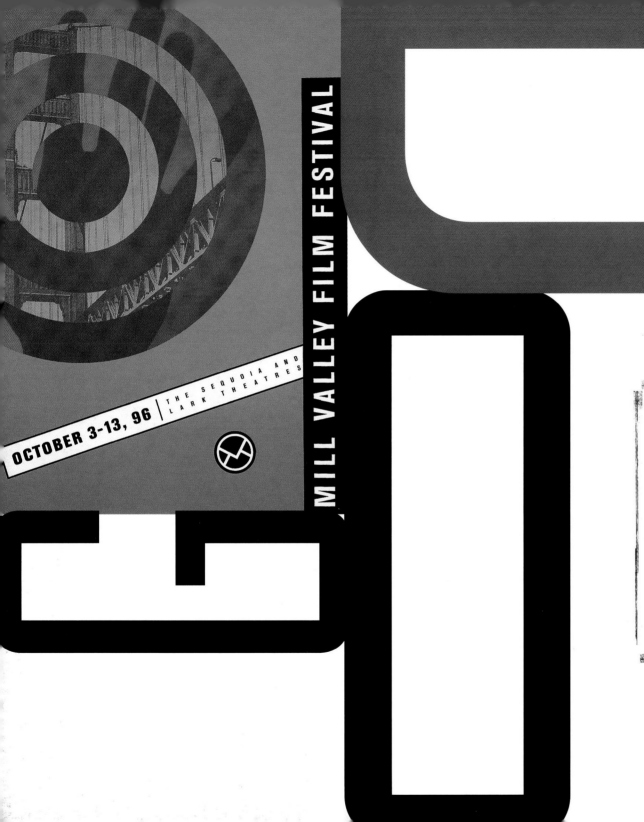

MARK FOX, a.k.a. BlackDog, came to digital design and illustration in 1986 from a UCLA degree in Fine Arts. The focus of his work there was photography and silk-screen, but on graduating he found that his mixed media background lent itself to design work. After opening his own studio in 1986 (and acquiring his first Macintosh in 1991) he now works primarily on creating logos and iconography for clients which include Texas soccer team Dallas Burn, Warner Bros. Records, Apple Computer, Aldus Corporation, Nike and Oracle. He also designs and markets a range of T-shirts under the name Bewear. Living and working in the San Francisco Bay area enables him to do a lot of design work for local cultural events, and the Mill Valley Film Festival work shown here was one such project.

EY FILM FESTIVAL

The Mill Valley Film Festival, now in its 19th year, is aimed at both film goers in the San Francisco Bay area and the movie industry. It premieres and promotes independent film, video and multimedia work from around the world with little funding, relying instead on sponsorship and volunteers. Fox was one of these, brought in to design a poster for the festival by ad agency Scheyer/sf, also working on the festival pro bono.

At Fox's suggestion, the one poster became a series of three, which were to be used as gifts for sponsors and as merchandising, with one litho-printed image being used as publicity material for the festival. 'In the past they've had good local talent design the posters: John Casado did them for years, and more recently Michael Mabry and Craig Frazier created images, and as far as a brief was concerned, it was nothing more than "design a great poster!",' enthuses Fox.

...odchenko, 'Lengiz' ('Advertising Poster for Books'). (Poster for Leningrad State Publishers). 1925.
Photography: Jim Frank and Merrill C. Berman Collection.

MILL VA

FINISHED WORK The final posters were silk-screened by hand and all colours custom-mixed. They now form part of the collection of architecture and design at the Museum of Modern Art in San Francisco.

Alexander Rodchenko, cover of "Novyilef" No 1. 1928.
Photography: Jim Frank and Merrill C. Berman Collection.

Jan Tschichold, 'Die Hose' (film poster). 1927.
Photography: Jim Frank and Merrill C. Berman Collection.

'I LOVED THE PURITY AND SIMPLICITY OF JAN TSCHICHOLD'S WORK, AND FOUND "DIE HOSE" VERY POWERFUL. EVEN THE USE OF TYPE AT AN ANGLE, WHICH I'VE ALWAYS SEEN AS AN AFFECTATION, INFLUENCED ME STRONGLY.'

INFLUENCES Fox found the way Jan Tschichold mixes flat areas of colour with photography in these film posters powerful, and also thought the angled type worked well: 'Tschichold's a dogmatic ass and he drives me crazy, but the purity and simplicity of these are quite nice,' he says.

CREATING DEPTH ⌐ 'These are typical of my work, and
while I like them, I felt they didn't
go far enough: they were too flat
and there was no tension
between the foreground and
background. The Mill Valley
posters were a direct result of my
dissatisfaction with the 5ive
Iconoclasts poster (far right), done
for the Design Lecture series
sponsored by the San Francisco
Museum of Modern Art,'
explains Fox.

(the project)

THE PROJECT
˅

From the outset, Fox had a few presuppositions about the project. 'I thought that I should use photography — film stills, in essence — for the primary imagery. Then, within the limitations of a static medium (the poster) I wanted to create an image that was in some way kinetic, that moved (like film),' he explains.

'I typically strive for narrative in my work, something which was obviously particularly relevent for this project, and I found that by pairing two photographs — regardless of their subject, or lack of subject — a story is immediately created,' he continues. 'Finally, I looked at Jan Tschichold's film posters and decided to let these influence my direction. This is evident primarily in the layout and handling of type,' he adds.

Needing a framing device, Fox decided to use what he describes as an obsession of his, the target or bull's-eye, which represents the eye and the camera lens, but also has a particular resonance for this project: 'Part of the logic behind using it here is its graphic similarity to tree rings, as Mill Valley, located in northern California, used to be a mill town and helped to build San Francisco through its logging operation,' he explains.

'AS MY LOGO WORK HAS TO BE **PERFECT**, DESIGNING POSTERS ALLOWS ME TO HAVE SOME FUN AND LOOK BEYOND STRAIGHTFORWARD COMMUNICATION TO ADDRESS AND INCORPORATE THINGS LIKE NARRATIVE AND **VISUAL RICHNESS**.'

final posters was shot and printed by Fox using an Olympus OM1. He scanned the photos for position only, before getting high-resolution scans from a bureau.

Sketches for the MV logo to brand Mill Valley. Inkings were then done before the image was realised in Illustrator.

(development)
DEVELOPMENTV

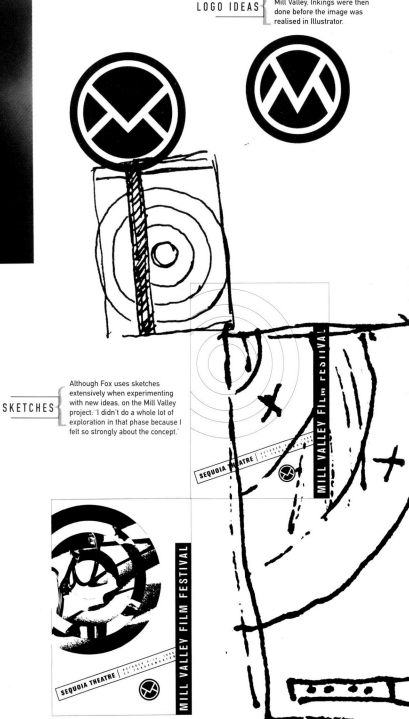

Having decided on the structure and elements of the posters, and feeling very strongly about the concept, Fox's thoughts developed in a fairly linear way, requiring little exploration or experimentation in the way of other directions. 'Much of the experimentation came when I shot the photos, as I had to decide which images would work/not work on their own, let alone when juxtaposed and forced into such a strange framing device as the target rings,' he says.

Getting the posters to reflect kineticism — the sense of movement that is integral to movie-making — was an important goal for Fox: 'I succeeded in creating an image that moves by juxtaposing two images in a target formation: the images move between the foreground and background; similarly, the viewer's eyes travel across the target's rings in an attempt to reconstruct missing elements from the photographs. I furthered this sense of movement by designing three posters using the same format and colour palette,' he explains.

SKETCHES   Although Fox uses sketches extensively when experimenting with new ideas, on the Mill Valley project: 'I didn't do a whole lot of exploration in that phase because I felt so strongly about the concept.'

While Fox had never been influenced by the work of Jan Tschichold in the past, this time he found inspiration in two Tschichold posters: 'On "Die Hose" and "Die Kameliendame" I loved the purity and simplicity of the work, and in "Die Hose" I particularly liked the mix of flat areas and colour three-dimensional imagery (the photography). It's very powerful, and even the use of type at an angle, which I've always seen as an affectation, influenced me strongly,' says Fox. And while Rodchenko's work is very different, his limited use of photography has parallels with Tschichold, but goes even further: In 'Advertising Poster for Books', from 1925, 'because the photographic image appears "real" in the context of the flat, geometric shapes which make up the poster, the viewer's eye is immediately pulled to it,' says Fox. 'It doesn't get any better than this,' he enthuses.

MILL VALLEY FILM FESTIVAL

SEQUOIA THEATRE | OCTOBER 3-13, 1996
25 THROCKMORTON

The basic design for the poster uses the bull's-eye, which reflects not only the camera lens and eye but also suggests the rings of a tree, which played a large part in Mill Valley's past of logging. For the subsequent comps (left), Fox used photos from the period between World War I and II; 'they helped set the direction for the poster series,' he says.

(technique) and KIT ∨

'I primarily design logos, so it was fun (albeit stressful!) to work on these posters,' says Fox. 'Creating them wasn't that different to creating a logo in the sense of having to get information across quickly and clearly, but they did allow me to create a richer, more complex narrative than I could achieve in a logo, which is why I enjoyed working on them so much,' he adds. And their success can be measured by their acceptance into the collection of architecture and design at the Museum of Modern Art in San Francisco.

Fox shot all the images used on the final posters and developed all the prints in a rented darkroom: 'It had been years since I worked in the darkroom!' He scanned the photos for position only and had a service bureau do the final scans at a higher resolution. 'The film was extremely expensive as the posters printed 24 x 36 inches full-bleed. They were silk-screened by hand, and all the colours custom-mixed. The monogram was hand-inked and then created in Adobe Illustrator 6. The individual elements were then brought together and the posters designed using Illustrator, then: 'I created the layout in Illustrator (placed the text, and so on) and dropped the fpo scans into it. Very straightforward. The scans were assembled as a series of masks in the target formation,' explains Fox.

EY FILM FESTIVAL

While the other two posters were done as silk-screens, this one was to also act as the publicity poster for the festival so was also reproduced as a litho print. While Fox would have preferred to use a human element such as an eye in the photography, in keeping with the use of lips and hand on the other two posters in the series, the festival organisers were particularly keen that he use the neon sign, which is that of a local restaurant.

SEQUOIA THEATRE | OCTOBER 3-13, 1996
25 THROCKMORTON

MILL VALLEY FILM FESTIVAL

'I WAS LOOKING AT EARLY BAUHAUS WORK AND COULDN'T FIGURE OUT WHAT TYPEFACES THEY WERE USING. SO I RESEARCHED IT AND FOUND AKZIDENZ GROTESK, PRODUCED BY BERTHOLD AS EARLY AS 1896! THE SANS SERIF MODERNISM OF THIS FACE AMAZES ME. I NEVER DETERMINED IF THE BAUHAUS BOYS USED IT, BUT I DID.'

HARDWARE

> 35MM OLYMPUS OM1 (FROM 1980)
> POWERMAC 8100/100
> HEWLETT PACKARD SCANJET IICX/T
> KOH-1-WOOR RAPIDOGRAPH

SOFTWARE

> ADOBE ILLUSTRATOR 6
> FUJI BLACK AND WHITE 400 ASA FILM

BLACKDOG
330 SIR FRANCIS DRAKE BOULEVARD, SUITE A
SAN ANSELMO, CA 94960, USA
TEL: +1 415 258 9663 FAX: +1 415 258 9681
E-MAIL: BlackDogma@aol.com

CLIENTS INCLUDE: San Francisco Museum of Modern Art; Warner Bros.
Records; PowerBar; Nike; Adobe Systems. [SEE PAGE 143]

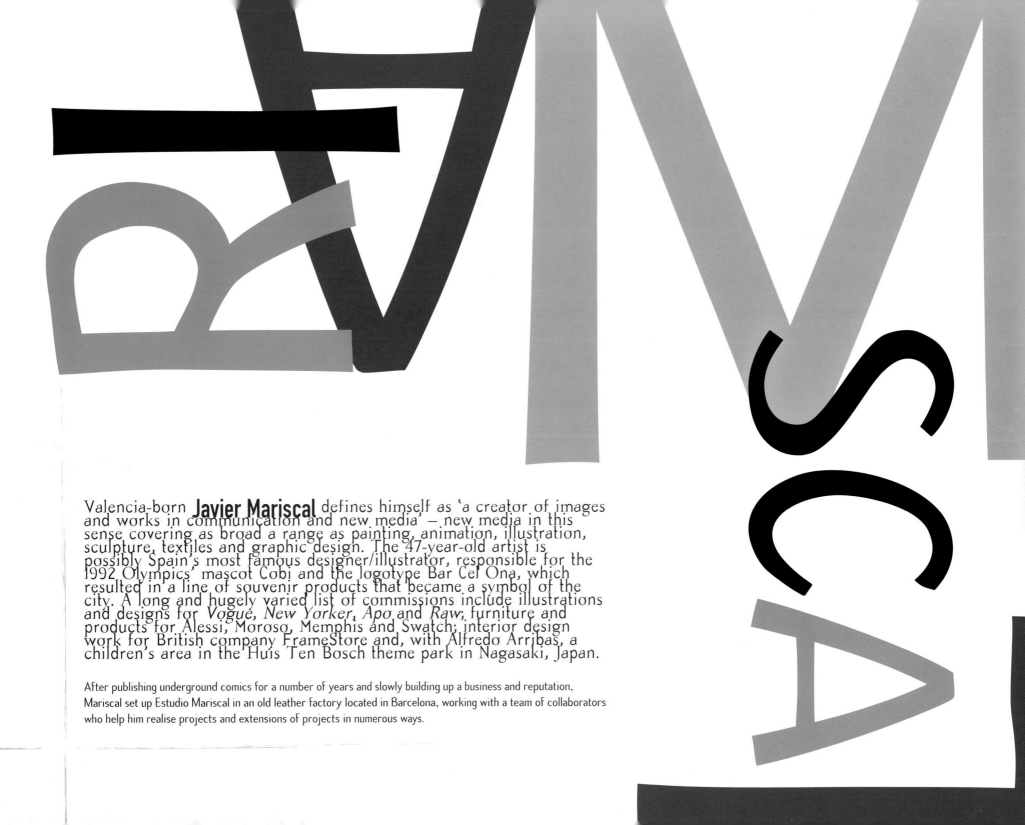

Valencia-born **Javier Mariscal** defines himself as 'a creator of images and works in communication and new media' — new media in this sense covering as broad a range as painting, animation, illustration, sculpture, textiles and graphic design. The 47-year-old artist is possibly Spain's most famous designer/illustrator, responsible for the 1992 Olympics' mascot Cobi and the logotype Bar Cel Ona, which resulted in a line of souvenir products that became a symbol of the city. A long and hugely varied list of commissions include illustrations and designs for *Vogue*, *New Yorker*, *Apo* and *Raw*; furniture and products for Alessi, Moroso, Memphis and Swatch; interior design work for British company FrameStore and, with Alfredo Arribas, a children's area in the Huis Ten Bosch theme park in Nagasaki, Japan.

After publishing underground comics for a number of years and slowly building up a business and reputation, Mariscal set up Estudio Mariscal in an old leather factory located in Barcelona, working with a team of collaborators who help him realise projects and extensions of projects in numerous ways.

(background)

BACKGROUND

∨

Last year Mariscal, along with 16 other international designers, was invited to enter a competition to design the mascot for the 2000 World Expo being held in Hanover. 'The overall presentation required an emotional image that symbolised the optimistic and human perspective of this World Exposition in a direct and sensuous way. And we had to come up with a character or mascot that could be further developed into a family of characters which was to be subsequently expanded,' says Mariscal of the competition.

The client's brief stressed: 'It is particularly important to bear in mind that it must be possible to execute the character and the family in three-dimensional form in a wide array of materials and sizes, and to ensure that it can function as a stand-alone, independent display.'

Mariscal and his team were chosen for the creation and implementation of the Expo's mascot and associated family of characters. 'Our mission was to develop and create a Corporate Design manual, which will be used as the guideline on how to implement and use the graphic material once the work is completed. We have also made a Corporate Design manual for animation and merchandising,' explains Mariscal.

(the project)

THE PROJECT

∧

THE BRIEF ┤ be designed in such a way that
they could be implemented in
two- and three-dimensional form,
in animated film, TV or computer
applications.'

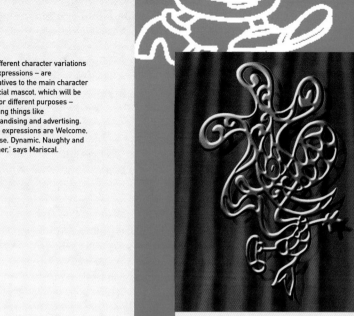

The different character variations
– or expressions – are
alternatives to the main character
or official mascot, which will be
used for different purposes –
including things like
merchandising and advertising.
'These expressions are Welcome,
Surprise, Dynamic, Naughty and
Dreamer,' says Mariscal.

VARIATIONS ┤

Twipsy

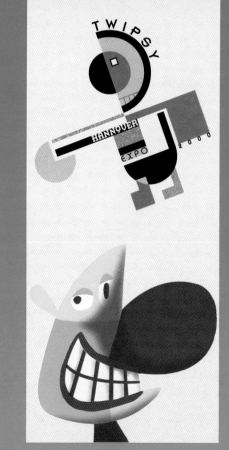

'THE INITIAL IDEA WAS THAT AT THE BEGINNING, A LONG TIME AGO,
A VERY LONG TIME AGO, THERE WAS A SUDDEN SPARK. A BANG. A
BIG BANG. A STAR WITH SHARP PRICKLES. A PIECE OF ENERGY
EXPANDING. A SHAPE THAT BECOMES ROUND.'

INTERPRETATION — The hundreds of different Twipsy realisations were the idea of Mariscal. The many interpretations and styles were 'accidental, intentional, and international,' says Mariscal.

DEVELOPMENT
(development)

Rather than discuss the technical or linear progression of Twipsy's realisation, Mariscal prefers to express the character's creation as he saw it initially, then developed it. What follows are those creative ideas behind Twipsy, who he is, why he is, and how he came into being.

'The initial idea was that at the beginning, a long time ago, a very long time ago, there was a sudden spark. A bang. A big bang. A star with sharp prickles. A piece of energy expanding. A shape that becomes round.'

'Something like a rock. An amoeba. Something aquatic. Something like a plant. Something that seems to have a mouth, a stomach, an eye. A bone structure. A digestive tract. An expression. A foot, a smile. An animal with long feet. Its tail could be a wing.'

'This animal smiles. Colours fill its body with expressions. Its legs move. A great arm appears. This animal is happy; jumps, shouts, dances, flies. It is blissfully happy.'

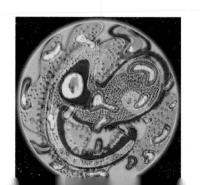

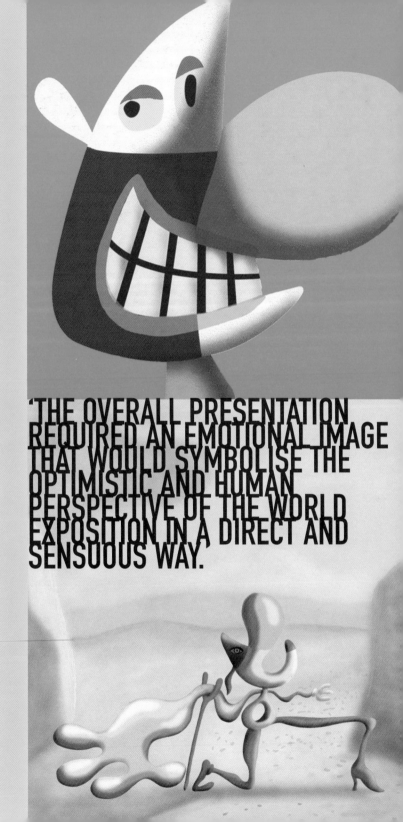

'THE OVERALL PRESENTATION REQUIRED AN EMOTIONAL IMAGE THAT WOULD SYMBOLISE THE OPTIMISTIC AND HUMAN PERSPECTIVE OF THE WORLD EXPOSITION IN A DIRECT AND SENSUOUS WAY.'

'We are in the year 2000 and there are still lots of things to be known, discovered and experienced. Twipsy is this smiling animal that wants to help us imagine the future. He has been developed as a puzzle. In pieces. Organic pieces, technological pieces and pieces of culture.'

'A very big mouth, an enormous nose, a very developed eye. The head of a very expressive and sensitive being. He has no problem making himself understood in any media. His body keeps on transforming. He is extremely alive. From the one-dimensional plane to the three-dimensional, he evolves and adapts himself to any situation.'

'He is an optimistic and funny friend and sometimes critical. He is emotional, sensitive and changes colour, skin, dress, shape according to his feelings and mood. A friend of every living creature. He is a traveller of outer space. A submarine navigator. A cybernetic explorer.'

TECHNIQUE

'One of Twipsy's arms is like a specialised vegetable or something very technological, and the other comes from the cartoon world. One foot is male and the other female.

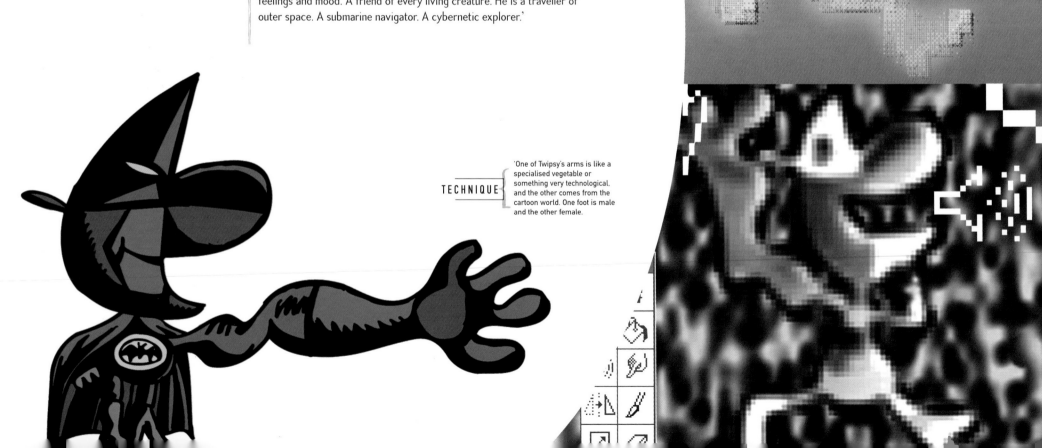

'IS HE AN ANIMAL, AN EXTRATERRESTRIAL OR A PIECE OF RAM? IS HE MADE OF PIXELS, OR FLESH AND BLOOD, OR IS HE A SPINACH THAT TALKS? HE IS TWIPSY. HE IS THE SOUL OF THE EXPO. THE HERO OF THE FUTURE.'

THEMES — Each set of figures – culture, science, technology, nature and ethnicity – relate to different aspects and themes which all play part of the Expo.

HARDWARE	SOFTWARE	
> APPLE POWER MACINTOSH 8500	> MACROMEDIA FREEHAND 5.5	> ADOBE STREAMLINE 3.0
> AGFA ARCUS II	> MACROMEDIA DIRECTOR	> ADOBE PHOTOSHOP 4.0
> IMPRESORA CANNON 700	> MACROMEDIA FONTOGRAPHER 4.0	> ADOBE SCREENREADY 1.0

JAVIER MARISCAL
ESTUDIO MARISCAL, PELLAIRES 30–38, 08019 BARCELONA, SPAIN
TEL: +34 93 303 34 20 FAX: +34 93 266 22 44
E-MAIL: estudio@mariscal.com
WEBSITE: http://www.mariscal.com

CLIENTS INCLUDE: 3M; Alessi; *El País*; Olympic Committee;
FrameStore; Swatch. [SEE PAGE 146]

GRANSHIP
Shizuoka Convention & Arts Center

zoo
Barcelona

THE LIGHTHOUSE
SCOTLAND'S CENTRE
FOR ARCHITECTURE
DESIGN AND THE CITY

GALLERY

ALESSANDRO BAVARI

VIA CARDUCCI 7, 04100 LATINA, ITALY
TEL: +39 773 696 828
E-MAIL: abArt@bmnet.it

Clients include: McKann Erikson; Romberg Contemporary Arts;
Fiumara d'Arte; WWF.

1	2	3	4
PERSONAL PROJECT USING PHOTOGRAPHY, PAINTS AND COMPUTER.	OREADI: COVER FOR GOTHIC-INDUSTRIAL MUSIC MAGAZINE *UTBR*.	RELIQUIARIO: COVER FOR GOTHIC-INDUSTRIAL MUSIC MAGAZINE *UTBR*.	CHIMERA: COVER FOR GOTHIC-INDUSTRIAL MUSIC MAGAZINE *UTBR*.

BLACKDOG

MARK FOX
330 SIR FRANCIS DRAKE BOULEVARD, SUITE A
SAN ANSELMO, CA 94960, USA
TEL: + 1 415 258 9663 FAX: +1 415 258 9681
E-MAIL: BlackDogma@aol.com

Clients include: San Francisco Museum of Modern Art; Warner Bros.
Records; PowerBar; Nike; Adobe Systems.

3

WILD BRAIN

4

1

2

5

1

LOGO/MONOGRAM FOR SAN FRANCISCO'S
MUSEUM OF MODERN ART EXHIBITION ENTITLED
'ICONS: MAGNETS OF MEANING'.

2

LOGO FOR THE SAN JOSE CLASH,
WORLD SOCCER LEAGUE.
(Art Director: Katy Tisch, Nike.)

3

LOGO FOR RED HERRING (A RETAIL STORE
IMPORTING FURNITURE AND HOUSEWARES IN
MILL VALLEY, CALIFORNIA).

4

WILD BRAIN, INC. LOGOTYPE (AN ANIMATION
STUDIO IN SAN FRANCISCO).

5

LOGO FOR ED CALDWELL, LOCATION
PHOTOGRAPHER FROM SAN FRANCISCO.

CLAUDIA NEWELL

484 CAPISIC STREET, PORTLAND, MAINE 04102, USA
TEL: +1 212 969 0795
E-MAIL: CNewell937@aol.com
WEBSITE: http://www.mindspring.com/~cnewell

Clients include: Nickelodeon; *The Houston Times*; *Los Angeles Times* magazine; Hemiola Records and *The Village Voice*.

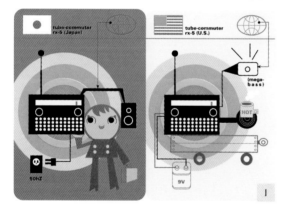

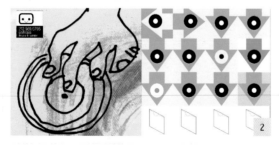

1

ILLUSTRATION FOR HIGH-END
HI-FI MAGAZINE *AUDIO*.

2

SELF-PROMOTIONAL SLEEVE FOR A
PORTFOLIO DISKETTE SHOWING MUSIC-
PACKAGING RELATED WORK.

3

ILLUSTRATIONS FOR *ADOBE* MAGAZINE ON NEW
FEATURES IN POSTSCRIPT 3.

4

A YEARLY 'BEST-OF' OF PRODUCTS AND
SERVICES IN THE COMMUNITY COVERED BY *THE
HOUSTON PRESS*, A WEEKLY PAPER.

DAVE McKEAN

FAX: +44 (0)1797 270 030

Clients include: DC Comics; Virgin Records; Saatchi & Saatchi US; Sony; *New Yorker*.

1	2	3	4	5
CD COVER FOR THE BEST OF BILL BUFORD'S EARTHWORKS COMPILATION. (Client: Virgin Records (Virgin Records).)	CD COVER FOR DOWNLOAD 111 BY CANADIAN BAND DOWNLOAD. (Client: Nettwerk Records, Toronto, Canada (Nettwerk Records).)	PHOTOGRAPH FROM MONOGRAPH OPTION:CLICK.	IMAGE FOR BOOK COVER FOR KODAK E100S FILM LAUNCH. (Client: Saatchi & Saatchi US (Kodak).)	*THE DAY I SWAPPED MY DAD* CHILDREN'S BOOK COVER. (Client: Neil Gaiman and Dave McKean.)

ESTUDIO MARISCAL

PELLAIRES 30–38, 08019 BARCELONA, SPAIN
TEL: + 34 93 303 34 20 FAX: +34 93 266 22 44
E-MAIL: estudio@mariscal.com
WEBSITE: http://www.mariscal.com

Clients include: 3M; Alessi; *El País*; Olympic Committee;
FrameStore; Swatch.

1

2

4

3

5

THE Lighthouse

SCOTLAND'S CENTRE
FOR ARCHITECTURE
DESIGN AND THE CITY

1 2	3	4	5
SILLÓN ALESSANDRA 1995: PART OF THE FIRST COLLECTION THAT MARISCAL PRESENTED AT THE MILAN FURNITURE FAIR IN ITALY (1995), 'LOS MUEBLES AMOROSOS DE MOROSO'. (Client: Moroso SpA.)	ZOO LOGO 1997: NEW GRAPHIC IMAGE FOR THE BARCELONA ZOO. (Client: Barcelona Zoo.)	GRANSHIP LOGO 1997–98: SHIZUOKA CONVENTION AND ARTS CENTER (JAPAN). (Architect: Arata Isuzaki. Project: Graphic Image, interior and exterior, signage.)	LIGHTHOUSE LOGO 1997: GRAPHIC IMAGE AND INTERIOR DESIGN FOR THE LIGHTHOUSE DESIGN AND ARCHITECTURE CENTRE IN AN OLD MACKINTOSH BUILDING IN DOWNTOWN GLASGOW.

FORK UNSTABLE MEDIA

JULIUSSTRASSE 25, 22769, HAMBURG, BUNDERSREPUBLIK
TEL: +49 40 432 948 12 FAX: +49 40 432 948 11
E-MAIL: da5d@fork.de E-MAIL: abbett.j@on-line.de

Clients include: Beiersdorf AG, b&d Verlag; Spar Deutschland;
Hamburger Sparkasse Astra Brauerei; Container Records, Hamburg;
Schultz & Friends, N.A.S.A., Hamburg.

1

3

5

2

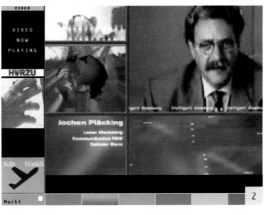

4

---

1 5

WEBSITE CONCEPT FOR TCHIBO COFFEE, A
LEADING GERMAN COFFEE DISTRIBUTOR AND
RETAILER. (Client: Schultz & Friends, N.A.S.A.
Hamburg.) The site was intended to be a freestyle and
ever-changing catalogue for hip lifestyle accessories as
well as travel-zine connected with the world of coffee and
supported through content-driven Shockware games.

2

A CD-ROM BASED PRESENTATION TOOL FOR
HIGHLIGHTING THE REDESIGN OF ONE OF THE
LEADING GERMAN TV MAGAZINES, *HOERZU*, 1998.
(Client: Popnet Kommunication, Hamburg.)

3

The magazine was looking for a new marketing
direction — younger audiences and progressive older
audiences. The main layout grid for the *Hoerzu*
presentation is based on 'test screen' which has a
colour band for each of the main presentation modules.

3 4

A VISUAL CONCEPT FOR MECURYBEATS
(POLYGRAM RECORDS), A HIP-HOP SITE ON THE
WEB. (Client: Rush Media, Hamburg.) The goal here
was to create a visually appropriate hip-hop site for
German netscape including a java-based graffiti/tag
section where users select different motives and can
share tags and artwork with others.

FOR ALL PROJECTS:

Creative Direction/Design: David Linderman.
Programming: Sascha Merg.
Project Management: Christian Schaumann/Manuel
Funk.

JASON STATTS

POST OFFICE BOX 16626, SAVANNAH
GEORGIA 31416, USA
TEL: +1 912 355 8398 FAX: +1 912 238 2459
E-MAIL: thinearth@worldnet.att.net

Clients include: *The Village Voice*; *Internet Underground* magazine;
The UTNE Reader; *Musician* magazine; *Jazziz* magazine; N Soul Records.

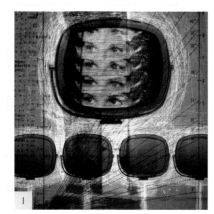

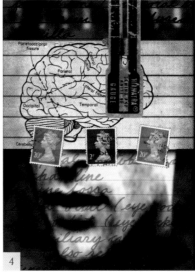

1	2	3	4	5
SADLY NOW DEFUNCT, *INTERNET UNDERGROUND* MAGAZINE COMMISSIONED STATTS ON A REGULAR BASIS.		PROMOTIONAL PIECE COMPLETED WHILE STATTS WAS A STUDENT UNDER PROFESSOR DON ROGERS.	MUTED COLOURS AND A MIX OF PHOTOGRAPHY, DIAGRAMS AND FOUND IMAGERY ARE A TRADE-MARK OF STATTS' DIGITAL AND TRADITIONAL ILLUSTRATION.	PUBLISHED IN SEPTEMBER 1997 FOR *JAZZIZ* ON-DISK MAGAZINE, WHICH PROMOTES NEW AND UP-AND-COMING JAZZ-ORIENTATED MUSIC.

JOHN J. HILL

12 JOHN STREET, SUITE 10, NEW YORK, NY 10038, USA
TEL: +1 212 766 8035
E-MAIL: jinn@pop.inch.com
52MM WEBSITE: http://www.inch.com/~jinn

Clients include: Acclaim Entertainment; Agency.com; American Express; Capitol Records; Virgin Entertainment; Flatline Comics; LaQuinta Filmworks; Axis Comics.

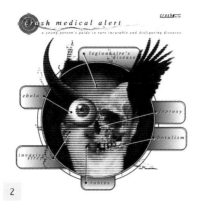

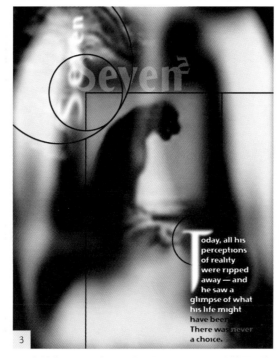

1

WEBSITEHOME
PROJECT: HAUS OF JINN HOMEPAGE
(http://www.inch.com/~jinn).
CREATED FOR: Self promotion.

2

CRASHSITE
PROJECT: CRASH MEDICAL ALERT
(http://www..crashsite.com/SofTV/Virus/).
CREATED FOR: Crash Site http://www.crashsite.com.
(Client: Big Gunn Project.)

3

SEVEN2
PROJECT: SEVEN2.
CREATED FOR: Personal study.

4

LEDGE
PROJECT: LEDGE.
CREATED FOR: 52MM promotional artwork.

5

SPAWN 1
PROJECT: SPAWN PIN-UP.
CREATED FOR: SPAWN#58 PIN-UP.
(Client: Todd McFarlane Productions.)
Spawn is ™ and © 1998 Todd McFarlane Productions.

J. OTTO SEIBOLD

1261 HOWARD STREET, #3
SAN FRANCISCO, CA 94103, USA
TEL: +1 415 558 9115 FAX: +1 415 558 9131
E-MAIL: jotto@pop.sirius.com (j. otto seibold)

Clients include: Jean Paul Getty Centre; Walker Art Centre.

1	2	3	4
CHILDREN'S BOOK FOR THE JOHN PAUL GETTY CENTER IN LOS ANGELES.	*OLIVE THE OTHER REINDEER*, CREATED BY J. OTTO SEIBOLD AND VIVIAN WALSH AND PUBLISHED BY CHRONICLE IN DECEMBER 1997.	PART OF AN ONGOING PROJECT WITH THE WALKER ART CENTER IN MINNEAPOLIS.	EXHIBITION POSTER FOR THE 'NEW POP: AMERICAN ILLUSTRATION' SHOW HELD AT THE PALAZZO DELLE ESPOSIZIONI IN ROME, 1996.

ME COMPANY

14 APOLLO STUDIOS, CHARLTON KINGS ROAD
LONDON NW5 2SA, UK
TEL: +44 (0)171 482 4262 FAX: +44 (0)171 284 0402
E-MAIL: meco@meco.demon.co.uk

Clients include: Nike; Coca-Cola; Warner Records; Polygram Records.

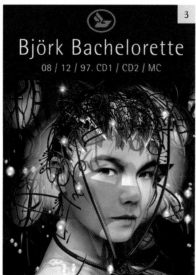

1	2	3	4	5
CHARACTER STICKER FOR 5 DOGS AD CAMPAIGN 1998.	UNIMAN FRONT UNIQ CHARACTER FOR UNILEVER GRADUATE RECRUITMENT WEBSITE.	BJÖRK 'BACHELORETTE' 30X20 POSTER. 1997. (Client: One Little Indian.)	BJÖRK 'HOMOGENIC' BOOKLET COVER. 1997. (Client: One Little Indian.)	BOKUS PROMOTIONAL IMAGE FOR BOKUS.COM. 1997.

PAUL SHORROCK

31 WYATT PARK ROAD, STREATHAM HILL, LONDON SW2, UK
TEL: +44 (0)181 674 4420 FAX: +44 (0)181 674 4427
E-MAIL: p.shorrock@mail.bogo.co.uk

Clients include: A&M Records; BBC; British Telecom; Channel 4; DDA;
DMB&B; EH6; *Financial Times*; IBM; The British Council; *Esquire*; *Time
Out*; *The Sunday Times*; *Daily* and *Sunday Telegraph*; Saatchi & Saatchi;
John Brown Publishing; *Time* Magazine; Lewis Moberly.

1

'LINE ONE'
COVER ILLUSTRATION FOR A SUPPLEMENT TO
THE SUNDAY TIMES ANNOUNCING THE LAUNCH
OF 'LINE ONE', AN INTERNET SERVICE CREATED BY
COLLABORATION BETWEEN BRITISH TELECOM-

2

'STRENGTH'
FRONT PAGE ILLUSTRATION FOR *FINANCIAL
TIMES* 'WEEKEND MONEY' SECTION ARTICLE
ABOUT CHOOSING LIFE ASSURANCE AND
PENSIONS.

3

'TV & RADIO'
ILLUSTRATION FOR *THE SUNDAY TIMES
MAGAZINE*'S '1000 MAKERS OF MUSIC' FOR AN
ARTICLE ABOUT THE ROLE PLAYED BY RADIO
AND TV SHOWCASES IN POPULAR MUSIC.

RIAN HUGHES

6 SALEM ROAD, LONDON W2, UK
TEL: +44 (0)171 221 9580 FAX: +44 (0)171 221 9589
E-MAIL: RianHughes@aol.com
WEBSITE: devicefonts.co.uk

Clients include: DC Comics; *Forbidden Planet*; MTV; Mambo;
Virgin Airlines.

1	2 3	4
FOONKY FONT POSTER: POSTER PROMOTING DEVICE FONTS FOONKY AND FOONKY STARRED.	YELLOW BOOTS NEW AND YELLOW BOOTS AUTUMN. IN-STORE DISPLAY HOARDINGS FOR JAPANESE CLOTHING CHAIN JUN CO. IN TOKYO. HUGHES ALSO DESIGNED MATCHING SWING TAGS AND BAGS.	ASSORTED DC COMIC LOGOS.

STEFAN SAGMEISTER

SAGMEISTER INC., 222 WEST 14TH STREET
NEW YORK, NY 10011, USA
TEL: +1 212 647 1789 FAX: +1 212 647 1788
E-MAIL: Ssagmeister@aol.com

Clients include: Graphics and packaging for David Byrne; Energy Records; Aerosmith; Warner Bros. Records; Lou Reed; HBO Studio Productions; Viacom.

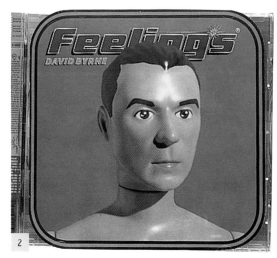

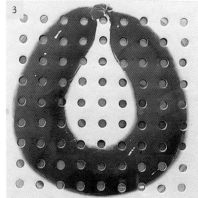

1	2	3	4
POSTER FOR THE NATIONAL CONFERENCE OF THE AMERICAN INSTITUTE OF GRAPHIC ART, HELD IN NEW ORLEANS.	CD PACKAGING FOR DAVID BYRNE'S 'FEELINGS' INCLUDES A 'MOOD COMPUTER' WITH WHICH YOU DETERMINE YOUR CURRENT FEELINGS.	CD PACKAGING FOR 'FANTASTIC SPIKES THROUGH BALLOONS' BY SKELETON KEY. FEATURES LYRICS WHICH HAVE BEEN REVERSED AS THE BAND DIDN'T WANT FANS READING THE LYRICS WHILE LISTENING TO THE MUSIC.	THE CALL FOR ENTRIES TO THE HONG KONG ADVERTISING AWARDS, ORGANISED BY THE 4 As (ASSOCIATION OF ACCREDITED ADVERTISING AGENCIES) FEATURES SAGMEISTER'S BEHIND DRAWN BY HONG KONG COPY PAINTERS. IT CAUSED A PREDICTABLE SCANDAL BUT PULLED IN MORE ENTRIES THAN BEFORE AND WON A GOLD AWARD.

STEVEN R. GILMORE

8702 CHALMERS DRIVE, LOS ANGELES, CA 90035, USA
TEL: + 1 310 659 4516 FAX: +1 310 659 4517
E-MAIL: srg@graphic.com
WEBSITE: www.webstorm.com/~srg

Clients include: EMI, The David Geffen Company; Warner Bros.; Polygram;
Industrial Light and Magic; Miramax; Coca-Cola.

1	2	3	4
POSTER ADVERTISING AIDS FUND-RAISER. (Client: With Pride.)	PARTY INVITE TO STEVEN R. GILMORE'S 39TH BIRTHDAY PARTY.	MAGAZINE AD FOR ARITZIA CLOTHING STORE.	ADVERTISING POSTER FOR 'GOOD WEIRD FEELING' ALBUM BY ODDS. (Client: Warner Music Canada.)

TERRY COLON

554 LOMBARD STREET, SAN FRANCISCO, CA 94133, USA
TEL: +1 415 276 8556/415 835 2422
E-MAIL: terry@www.suck.com

Clients include: *Fortean Times*; Visable Ink Press; *Cracked* Magazine;
Applause!; Unix World; Home Office Computing; Ford Motor Company;
Infoworld Magazine; Wired Ventures.

3

2

1

1

KING KONG IN HOLLYWOOD FOR
FORTEAN TIMES.

2

COVER OF THE *SUCK* BOOK, PUBLISHED BY
WIRED VENTURES.

3

SCALE-A-TRONIC FOR *INFOWORLD* MAGAZINE.

TYPERWARE®

ANDREU BALIUS AND JOAN CARLES P. CASASÍN
ANSELM CLAVÉ, 11, 08106-SANTA MARIA DE MARTORELLES,
BARCELONA, SPAIN TEL/FAX: +34 93 579 09 59
E-MAIL: typerware@seker.es WEBSITE: http://bbs.seker.es/~joanca

Clients include: Fundació Miró; La Fura dels Baus; Editorial Anagrama;
Accidents Polipoétics; Fundació La Caixa; Sol Picó Dance Company;
Museu Nacional d'Història de Catalunya.

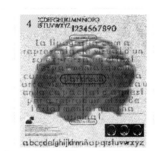

1
PRIMAVERA SOUND
POSTER ADVERTISING POP CONCERT FOR MTG.

2
JEDI KNIGHTS
POSTER ADVERTISING GIG AND DJING AT
BARCELONA DANCE CLUB NITSA CLUB BY ELEC-
TROMUSIC BAND JEDI KNIGHTS.
　(Client: MTG (MurmurTownLaGlòria Productions.)

3
GARCIA AND CO. FONT CATALOGUES
FONT CATALOGUES DESIGNED BY ANDREU
BALIUS FOR A DIVISION OF TYPERWARE®,
GARCIA FONTS.

4
INSPECTOR TUPPENCE
LAUNCH PARTY POSTER FOR S/T, A FANZINE
INCLUDED IN THE *AJOBLANCO* MAGAZINE.

ACKNOWLEDGEMENTS

ACKNOWLEDGEMENTS

HUGE THANKS MUST GO TO ALL WHO CONTRIBUTED TO THIS BOOK, FOR THEIR WILLINGNESS TO PARTICIPATE IN WHAT WAS VERY MUCH A COLLABORATIVE AND LABOUR-INTENSIVE PROJECT. THEIR EFFORT AND ENTHUSIASM NEVER FALTERED AND THEY WERE ALWAYS HELPFUL AND PATIENT. THE SAME GOES FOR ALL THOSE INVOLVED IN PUTTING THE BOOK TOGETHER, INCLUDING NATALIA PRICE-CABRERA AT ROTOVISION AND MARK ROBERTS AND MATT POWELL AT THE DESIGN REVOLUTION. COLLEAGUES AND MANY FRIENDS WHO SAW ME THROUGH THE LONG DAYS WITH A SYMPATHETIC 'HOW'S THE BOOK GOING?' THEN LISTENED PATIENTLY AS I TOLD THEM — AT LENGTH — DESERVE MY GRATITUDE, AS DOES JOHN CRANMER, WITHOUT WHOM THE DAVE McKEAN SECTION OF THE BOOK WOULD NOT HAVE BEEN POSSIBLE, AND PAUL MURPHY, WITHOUT WHOM THE REST OF IT WOULD NOT HAVE BEEN POSSIBLE. FINALLY, THE RABBIT FOUNDATION WILL ALWAYS BE REMEMBERED FOR ITS SUPPORT.